Large Language Models to Streamline Your Everyday Business

Accelerate Efficiency, Drive Innovation, and Maximize ROI

Elizabeth Scalzo

CONTENTS

Introduction

"I have great hopes in this direction for machines that will rival or even surpass the human brain"

– Claude Shannon, the father of information theory

All too often today, businesses find it difficult to keep up with the pace of technology. Despite a talented workforce and innovative products, companies struggle to keep pace with the breakneck speed of the digital age. That is, until they discover the transformative power of Large Language Models (LLMs). By harnessing the potential of these advanced AI systems, companies can not only streamline their processes but also unlock new avenues for growth and innovation.

This book aims to demystify the world of LLMs for business leaders like you, providing a clear roadmap for successful AI integration into your organization. As the pace of technological change accelerates, understanding and leveraging LLMs has become a critical factor in driving innovation and maintaining a competitive edge. Whether you're a CEO of a small startup or an executive at a multinational corporation, this book will equip you with the knowledge and tools necessary to set the direction of your organization toward making the best use of AI to transform your business.

As a software engineer with over three decades of experience, I have witnessed firsthand the profound impact of AI on businesses across industries. My expertise in system architecture design and Independent Verification and Validation has given me a unique perspective on the challenges and opportunities that come with integrating advanced technologies like LLMs into existing business processes. Through this book, I aim to bridge the gap between the technical intricacies of LLMs and their practical applications in a business context.

The book is structured to cater to readers with different levels of expertise and interest. Those seeking a concise overview can focus on the first three chapters, which provide a high-level introduction to LLMs, their evolution, and their potential impact on businesses. Readers looking for a more comprehensive understanding can read the subsequent chapters, which cover technical terminology, implementation strategies, ethical considerations, and future trends in greater detail. For those seeking an in-depth exploration of the subject, the entire book offers a wealth of information and insights.

Throughout the book, you'll encounter real-world case studies and examples that illustrate the transformative power of LLMs in action. From streamlining customer service to optimizing supply chain management, these examples will inspire you to think creatively about how LLMs can be applied to your own business challenges. You'll also gain a deeper understanding of the ethical considerations surrounding AI implementation, ensuring that your organization's use of LLMs aligns with your values and social responsibilities.

By the end of this book, you'll have a clear understanding of the key concepts and technologies behind LLMs, as well as a practical framework for integrating them into your business processes. You'll be equipped with the knowledge and tools necessary to make informed decisions about AI adoption, calculate the return on investment, and build a compelling business case for LLM implementation. Most importantly, you'll be prepared to lead your organization into the future, harnessing the power of LLMs to drive innovation, efficiency, and growth.

As we embark on this journey together, I invite you to approach the subject with an open mind and a willingness to embrace change. The world of AI is rapidly evolving, and staying ahead of the curve requires a commitment to continuous learning and adaptation. By investing in your understanding of LLMs and their potential applications, you're not only positioning your business for success but also contributing to the broader conversation about the role of AI in shaping our collective future.

So, let us begin by exploring the fascinating history and evolution of Large Language Models, setting the stage for a deep dive into their transformative potential for businesses like yours.

Chapter One:
A Short History Of AI

"We can build a much brighter future where humans are relieved of menial work using AI capabilities."

– Andrew Ng, CEO of DeepLearning.AI

In the busy corridors of a multinational conglomerate, a team of data scientists huddled around their computers, eyes fixed on screens displaying lines of code that promised to revolutionize their customer engagement strategy. The stakes were high; they were deploying a new system, the latest generation of AI technology—Large Language Models (LLMs)—to enhance customer service operations. After they finished fine-tuning the parameters, the LLM began processing vast swaths of data, generating insights and responses with a level of sophistication that had previously been unimaginable. This new system represented a new era in how businesses understood and interacted with their clientele.

This chapter charts the evolution of LLMs from their theoretical origins to their current state as indispensable tools in modern business. By tracing the roots of these models, we can appreciate the many breakthroughs that have brought us to this point, laying the groundwork for the strategic deployment of LLMs across various industries. Understanding this history is essential for any business leader seeking to harness the transformative power of AI, as it provides context for both the capabilities and limitations of these formidable tools.

1. From Theory to Practice: The Rise of LLMs

The saga of Large Language Models began in the mid-twentieth century with the emerging field of artificial intelligence. This was a time when pioneers like Alan Turing and Claude Shannon laid the foundational theories for processing natural human language. A pivotal figure at this time was Noam Chomsky, whose theory of universal grammar posited that the human brain comes pre-equipped with a set of linguistic principles shared across all languages. This paradigm shift challenged the prevailing behaviorist views of language acquisition, which focused on external stimuli and reinforcement. Chomsky introduced the concept of a Language Acquisition Device (LAD), a theoretical construct in the brain that facilitates the rapid learning of language. Although primarily a linguistic theory, Chomsky's work provided an intellectual framework that would influence the development of computational models of language, as it suggested that language processing—whether by humans or machines—relies on an inherent understanding of structural rules. This concept would become central to AI research in subsequent decades.

The transition from theory to tangible AI applications came with the advent of neural networks in the latter half of the twentieth century. Early AI systems were rule-based, relying on explicitly programmed instructions to process language. These systems faced significant limitations, particularly their inability to handle the ambiguity and variability inherent in natural human communication. The emergence of neural networks marked a paradigm shift toward data-driven approaches, where machines could learn patterns from large datasets rather than relying solely on predefined rules. Data-driven software was a revolutionary development catalyzed by advances in computational power and the availability of increasingly large datasets.

In the 21st century, AI systems have witnessed rapid technological advancements. The development of deep learning algorithms, particularly those using multi-layered neural networks, enabled machines to process and understand language with unprecedented accuracy. This era also saw the introduction of backpropagation—a method for training neural networks by adjusting weights based on error rates—which allowed models to produce results very similar to what humans might produce. The combination of these innovations set the stage for Large Language Models capable of tackling complex natural language processing tasks with efficiency and precision.

As we moved beyond rule-based systems, probabilistic models emerged, enabling machines to make predictions based on probabilities derived from data patterns. This shift was significant because it allowed for handling uncertainty and variability in language, elements with which rule-based systems struggled. The move towards machine learning and data-centric processes further empowered LLMs to adapt and improve over time, learning from vast amounts of data to enhance their performance continuously. This adaptability makes LLMs very powerful in modern applications, as they can be fine-tuned for a wide array of tasks, from sentiment analysis to language translation.

In the early stages of their deployment, LLMs found applications across various industries, revolutionizing the way businesses approached language-related tasks. In customer service, for example, LLMs were employed to automate responses in call centers, reducing wait times and improving customer satisfaction. The field of natural language processing (NLP) was a natural fit for these models, as they could analyze and generate human-like text, making them invaluable for tasks such as generating reports or summarizing documents. Initial experiments in language translation also showcased the potential of LLMs to break down language barriers, enabling seamless communication across different linguistic communities.

Case Study: Early Applications of LLMs

One of the earliest adopters of LLM technology was a global telecommunications company facing mounting pressure to improve customer service without incurring prohibitive costs. By implementing LLMs in their call centers, they achieved a 30% reduction in response times and a 20% increase in customer satisfaction scores. The LLMs were trained to understand the nuances of customer inquiries, allowing them to provide accurate and contextually relevant responses, thereby freeing up human agents to handle more complex queries. This case study illustrates the practical benefits of LLMs in automating routine tasks while enhancing overall service quality.

The transition from theory to practice in the field of LLMs has been a journey marked by groundbreaking innovations and transformative applications. As you learn more about the capabilities of these models, it becomes clear that their potential extends far beyond mere automation; they represent a paradigm shift in how businesses can engage with language, data, and ultimately, their customers.

2. Key Milestones in LLM Evolution

In the ever-evolving field of artificial intelligence, the development of Large Language Models (LLMs) created a series of pivotal technological breakthroughs, each representing a significant leap forward in our quest to replicate human-like understanding and language generation in machines. Among these, Vaswani et al.'s introduction of the Transformer model in 2017 stands as a watershed moment. This model departed from previous architectures by employing a mechanism known as self-attention, which allowed it to weigh the significance of different words in a sentence regardless of their position, thereby capturing more context and nuance than ever before. The Transformer model's ability to process language in parallel rather than sequentially dramatically increased its efficiency and scalability, laying the groundwork for the subsequent development of more sophisticated LLMs like BERT (Bidirectional Encoder Representations from Transformers) and GPT-3 (Generative Pre-trained Transformer 3).

BERT, introduced by Google in 2019, marked another milestone in the evolution of LLMs by setting a new benchmark for performance in natural language processing tasks. Its bidirectional approach allowed for a more holistic understanding of context by considering the entire sentence rather than just the preceding or following words. This capability was groundbreaking for tasks that required nuanced comprehension, such as sentiment analysis and question answering. The impact of BERT was profound, as it enabled a level of linguistic understanding that was previously unattainable, thus enhancing the precision and accuracy of AI-driven language applications across many kinds of domains.

Similarly, the advent of GPT-3 by OpenAI in 2020 represented an unprecedented leap in language generation capabilities. With an astonishing 175 billion parameters, GPT-3 demonstrated a remarkable ability to generate coherent, contextually relevant text that closely mimicked human writing. Its prowess in producing text that could pass as human-written opened new vistas for creative applications, from content creation to automated dialogue systems. The implications of GPT-3's capabilities were vast, prompting businesses to reimagine how they could leverage AI to augment human creativity and productivity in ways they had always thought were the exclusive domain of human intellect.

Contributions from open-source communities have significantly accelerated LLMs' rapid evolution and democratized access to cutting-edge AI technologies. Initiatives by organizations like OpenAI and Google AI have played crucial roles in this democratization process. OpenAI's decision to make models like GPT-2 accessible to the public was a catalyst for widespread experimentation and innovation. Collaborative platforms such as GitHub have further facilitated this process by providing a space for researchers and developers to share code, insights, and improvements. This collective effort has led to the continuous iteration and refinement of LLM models, ensuring that advancements are not confined to a select few but are instead accessible to a global community of innovators.

The scaling of LLMs has been another critical factor in enhancing their capabilities because of the extensive hardware required and the gigantic datasets on which they are trained. The exponential growth in dataset sizes, driven by the digital proliferation of data, has provided LLMs with the vast amounts of information necessary to develop nuanced language models; however, this scaling has not been without its challenges. Training large-scale models necessitates significant computational resources, tests the limits of Moore's Law, and necessitates innovative solutions to manage the increased complexity and cost. Techniques, such as distributed training, and specialized hardware, like GPUs and TPUs, have emerged as solutions to these challenges and have enabled LLMs' continued growth and sophistication.

As LLMs have become more widespread, they have also attracted increased scrutiny regarding regulatory and ethical considerations. The introduction of AI ethics guidelines has been a pivotal development, providing a framework for addressing issues related to transparency, accountability, and fairness in AI systems. Regulatory bodies worldwide have begun to respond to the rapid advancements in AI with guidelines and legislation to ensure that the deployment of LLMs does not exacerbate existing biases or inequalities. As stakeholders seek to mitigate the risks of perpetuating harmful stereotypes or making decisions that could negatively impact marginalized communities, the discussion of bias and fairness has become central to the conversation around LLMs.[1]

The journey of LLMs from theoretical constructs to powerful tools in the modern technological arsenal is a testament to the

---

[1] See UNESCO, "Ethics of Artificial Intelligence".

relentless pursuit of innovation and the collective efforts of a global community of researchers, developers, and thought leaders. As we continue to explore the possibilities and challenges presented by these models, it is clear that the milestones we have achieved thus far are merely the beginning of a much larger story in the evolving field of artificial intelligence.

3. LLMs Boost Business Potential in Today's World

Large Language Models (LLMs) have emerged as transformative instruments today, driving unprecedented levels of innovation and operational efficiency across many sectors. These sophisticated AI tools redefine how businesses perceive and engage with language, enabling them to extract deeper insights and deliver more personalized experiences to their customers. Many companies have dabbled with ChatGPT or Claude 3.7 just to enhance their everyday tasks; however, they are missing much of the value of LLMs unless they sign up for a deeper dive. For example, in the financial sector, LLMs are being leveraged to analyze and interpret vast amounts of textual data from sources such as news articles, financial reports, and social media. This capability allows financial analysts to identify trends, assess risks, and make informed investment decisions with a speed and accuracy that was previously unattainable. Moreover, LLMs are facilitating the automation of routine tasks such as report generation and compliance checks, freeing up human analysts to focus on more strategic initiatives.

In the healthcare industry, the impact of LLMs is equally profound. These models are being used to sift through mountains of medical literature and patient records to assist in diagnosing complex conditions, recommending personalized treatment plans, and even predicting patient outcomes. By processing natural language data from clinical notes and research papers, LLMs can identify patterns and correlations that might elude even the most experienced medical professionals. This analysis not only enhances the quality of care but also accelerates the pace of medical research, leading to the development of innovative therapies and interventions. Furthermore, LLMs play a pivotal role in streamlining administrative processes within healthcare organizations, reducing the burden of paperwork, and improving overall efficiency.

In customer service, LLMs are revolutionizing the way businesses interact with their customers. By powering intelligent chatbots and virtual assistants, these models enable organizations to provide round-the-clock support, addressing customer queries and resolving issues with remarkable accuracy and empathy. This customer support not only enhances customer satisfaction but also reduces the workload on human customer service representatives, allowing them to focus on more complex and nuanced interactions. The ability of LLMs to understand and generate human-like text also enables businesses to personalize their communication, tailoring messages to the individual preferences and needs of their customers. This level of personalization fosters stronger customer relationships and drives brand loyalty, ultimately contributing to increased revenue and market share.

The retail sector, too, is witnessing a paradigm shift with the adoption of LLMs. These models enhance the shopping experience by providing personalized product recommendations, optimizing inventory management, and predicting consumer behavior. LLMs can identify emerging trends and preferences by analyzing customer reviews, social media interactions, and historical sales data, enabling retailers to make data-driven decisions about product offerings and marketing strategies. This analysis not only improves customer satisfaction but also boosts sales and profitability. In addition, LLMs automate content creation for product descriptions, marketing copy, and social media posts, ensuring consistency and coherence across all customer touch-points.

LLMs are streamlining the research and analysis process in the legal industry by automating the review of legal documents, contracts, and case law. This automation allows legal professionals to focus on the strategic aspects of their work, such as crafting arguments and advising clients, rather than getting bogged down in the minutiae of document review. The ability of LLMs to quickly and accurately analyze large volumes of text is particularly valuable in due diligence processes, where time is of the essence. Moreover, LLMs can assist in drafting legal documents by generating templates and suggesting language, thus improving efficiency and reducing the risk of errors.

Transportation and logistics companies are also reaping the benefits of LLMs. LLMs can optimize routing and scheduling by analyzing data from various sources, including shipment records, traffic reports, and weather forecasts, ensuring that goods are delivered on time and at the lowest possible cost. This enhances operational efficiency and improves customer satisfaction, as deliveries are more reliable and predictable. Furthermore, LLMs can predict maintenance needs for vehicles and equipment, reducing downtime and extending the lifespan of assets.

LLMs personalize learning experiences in education by analyzing student performance data and providing tailored recommendations for study materials and learning paths. This analysis allows educators to identify areas where students may be struggling and to provide targeted support, enhancing learning outcomes. LLMs can also automate administrative tasks such as grading and feedback, freeing educators to focus on teaching and mentoring.

The potential of LLMs extends beyond these examples, with applications in fields ranging from agriculture to energy, media, and beyond. In agriculture, LLMs analyze crop data and weather patterns, enabling farmers to make informed decisions about planting and harvesting. In the energy sector, LLMs are optimizing the operation of power grids by predicting demand and ensuring efficient distribution. In media and entertainment, LLMs generate scripts, compose music, and even create visual content, pushing the boundaries of creativity and innovation.

The finance industry is a leading adopter of LLMs, for example, BloombergGPT, a 50 billion parameter model, is specifically designed for financial tasks such as sentiment analysis and question answering. Typically, you'll see applications in financial analysis, risk assessment, and investment advice.

The key benefits of LLMs across these diverse sectors include increased efficiency, enhanced decision-making, improved customer experiences, and the ability to uncover insights and opportunities that were previously hidden. By automating routine tasks and enabling more sophisticated analysis, LLMs empower businesses to focus on what truly matters: driving innovation, delivering value to customers, and achieving strategic goals. As these models evolve, their potential to transform industries and redefine business paradigms will only grow, making them an indispensable tool for any organization looking to thrive in the digital age.

4. Use Cases for Large Language Models (LLMs)

In the vibrant ecosystem of modern business, the deployment of Large Language Models (LLMs) has become an instrumental force, driving transformation across diverse sectors. The adaptability of LLMs allows them to be tailored to the specific needs and challenges faced by different industries, offering bespoke solutions that enhance efficiency, creativity, and strategic decision-making. Let us explore how leading enterprises capitalize on these models to address real-world challenges and unlock new opportunities.

In finance, LLMs are indispensable tools for risk management and fraud detection. By processing and analyzing vast quantities of financial data, these models can identify anomalies and patterns that may indicate fraudulent activity, allowing institutions to take proactive measures to protect their assets and maintain the integrity of their operations.

Additionally, LLMs generate comprehensive financial reports, offering insights and forecasts that aid in strategic decision-making. This capability allows financial professionals to focus on crafting innovative investment strategies rather than being bogged down by time-consuming data analysis.

The media industry is harnessing the power of LLMs to streamline content creation and distribution. These models are used to generate articles, summaries, and headlines, producing high-quality content at a rapid pace. By analyzing trends and audience preferences, LLMs help media companies tailor their content to appeal to target demographics, increasing engagement and driving revenue. Additionally, LLMs facilitate the translation and localization of content, allowing media organizations to expand their reach and cater to global audiences, breaking down linguistic barriers that previously limited their scope.

In the manufacturing sector, LLMs optimize supply chain management and production processes. By analyzing data from sensors, machines, and logistics systems, these models can predict equipment failures, schedule maintenance, and optimize production schedules, reducing downtime and enhancing operational efficiency. This predictive capability allows manufacturers to maintain high levels of productivity and quality, ensuring they remain competitive in a rapidly evolving industry.

Across these diverse sectors, the common thread is the ability of LLMs to process vast quantities of data, extract meaningful insights, and automate complex processes. The versatility and adaptability of these models make them powerful allies in the quest for innovation and efficiency, enabling organizations to navigate the challenges of the modern business environment with confidence and agility. As businesses continue to explore and implement LLMs, the potential for transformative impact becomes increasingly apparent, promising a future where AI-driven solutions are integral to success.

Overall, businesses are reporting that they are seeing multiple benefits from integrating AI tools into their work. They are gaining insights into their data which enables data-driven decisions. AI enables work to proceed at a speed that is beyond human ability by automating mundane, repetitive tasks. Automating these repetitive tasks allows humans to focus on other tasks that require a higher level of judgment. AI is helping to support customers online through chatbots, and firms like Netflix are using AI to create more personalized recommendations. Manufacturing is using AI to monitor their systems and improve quality control. Financial organizations are seeing error-free reconciliations and detecting more frauds. Human Resource departments are using AI to facilitate their hiring process, remove bias in corporate communication, and gauge employee sentiment to retain the best employees. Organizations are seeing more innovation, higher profits, and new gains as more is learned about emerging capabilities. [2]

12 important benefits of AI for business

1. Better decisions
2. Efficiency and productivity gains
3. Improved speed of business
4. New capabilities and business model expansion
5. Personalized customer services and experiences
6. Improved services
7. Improved monitoring
8. Better quality and reduction of human error
9. Better talent management
10. More innovation
11. Increased profitability
12. Industry-specific improvements

As we conclude our exploration of the many applications of LLMs, it becomes clear that these models represent more than just a technological advancement; they are catalysts for change, driving innovation and fostering new ways of thinking across industries. By embracing the capabilities of LLMs, businesses can position themselves at the forefront of

---

[2] See TechTarget, "12 Key Benefits of AI for Business".

progress, leading the charge in creating a more interconnected, efficient, and intelligent world.

As we move forward, the question is not whether to adopt LLMs, but how to harness their full potential to achieve strategic objectives and deliver unparalleled value to customers.

Chapter Two:
Understanding Basic Terminology

"Learning and innovation go hand in hand. The arrogance of success is to think that what you did yesterday will be sufficient for tomorrow."

—William Pollard, American physicist and Episcopal priest

In the headquarters of a busy global corporation, imagine a scenario where seasoned executives gather around a boardroom table, their laptops open and pens poised, ready to decipher the language of the future. A language not spoken by humans but by machines—artificial intelligence, specifically Large Language Models (LLMs)—has become imperative for strategic decision-making. Yet, the intricacies of this language remain elusive to many, shrouded in technical jargon that often alienates many business leaders. This chapter will translate the complex lexicon of AI into terms that are not only comprehensible but also actionable, empowering you to leverage these technologies with confidence and precision.

1. Core Concepts

At the core of understanding LLMs lies the distinction between Artificial Intelligence (AI) and Machine Learning (ML), two terms often used interchangeably but fundamentally distinct in scope. AI encompasses the overarching ambition to replicate human intelligence in machines, enabling them to perform tasks that traditionally require human cognition. It is the grand vision of giving machines reasoning, perception, and decision-making capabilities. Machine Learning, on the other hand, is a subset of AI that focuses on the development of algorithms that allow systems to learn and improve from experience without being

17

explicitly programmed. It is the engine that powers Generative AI, which creates new content like text, images, videos, and music and drives innovations in areas ranging from predictive analytics to autonomous vehicles. It uses machine learning models to learn patterns from data and then create new content based on those patterns.

Natural Language Processing (NLP) represents another pivotal concept that bridges human language and machine understanding. NLP encompasses the methodologies and technologies that enable machines to comprehend, interpret, and generate human language in a meaningful and contextually appropriate manner. It is the cornerstone upon which LLMs are built, providing the foundation for applications such as chatbots, sentiment analysis, and language translation. Within Generative AI, LLMs are particularly adept at producing human-like text, leveraging deep neural networks to predict and generate language with remarkable fluency.

Deep learning, the architectural marvel underpinning LLMs, is a model that employs neural networks with multiple layers to simulate the workings of the human brain. Just as our brain has a complex network of neurons that fire signals along pathways to represent our thought processes, the LLM neural network consists of interconnected nodes that process information through a series of weighted connections that allow the neural network to learn complex patterns from vast datasets. This analogy provides a tangible framework for understanding how deep learning models can achieve such sophisticated levels of comprehension and generation.

As we learn more jargon associated with LLMs, the term "token" emerges as a fundamental building block. Tokens are the discrete units into which text is broken down for processing by the model, much like you parse words or phrases to help you understand human language. The model then transforms these tokens into "embeddings", numerical representations that capture the semantic and syntactic nuances of the language. The embeddings enable the model to understand context and meaning. Attention mechanisms further enhance this understanding by allowing the model to focus on the most relevant parts of the input, much like how you might hone in on key points of a conversation. This focus is critical for tasks requiring nuanced interpretation, such as question-answering or language translation.

Training data and model parameters are other essential components in the lexicon of LLMs. Training data refers to the vast corpora of text used to teach the model, providing the examples from which it learns patterns and structures. Model parameters, on the other hand, are the adjustable weights and biases that determine how the model processes and generates language. These parameters are fine-tuned during the training process to optimize performance and accuracy. Pre-training and fine-tuning are two phases in the lifecycle of an LLM, where pre-training involves exposing the model to a broad dataset to develop general language understanding, and fine-tuning tailors the model to specific tasks or domains by training it on targeted data.

To elucidate these concepts further, consider the analogy of a neural network as a vast network of roads and highways, each neuron representing an intersection where information converges and diverges. As data flows through this network, it is like vehicles traveling along these routes, with the weights and biases functioning as traffic signals that control the flow and direction of information. Just as a well-designed transportation system facilitates efficient movement, a well-calibrated neural network enables the seamless processing of complex language tasks.

In practical business applications, these technical terms translate into tangible benefits. For instance, AI-driven chatbots powered by NLP can enhance customer interaction by providing instant, accurate responses to inquiries, thereby improving customer satisfaction and operational efficiency. In the retail sector, deep learning models can refine image recognition systems, enabling more effective inventory management and personalized shopping experiences.

Similarly, NLP applications in sentiment analysis can offer valuable insights into consumer opinions and market trends, informing strategic marketing decisions.

Interactive Element: AI Jargon Quiz

To reinforce your understanding of these concepts, consider engaging in the following interactive quiz designed to test your grasp of AI terminologies and their applications. This exercise will not only solidify your knowledge but also provide a platform for reflection on how these terms can be applied within your own organization:

1. What is the primary distinction between AI and ML?

2. How do tokens contribute to the processing capabilities of an LLM?

3. Why are attention mechanisms crucial in NLP tasks?

4. In what ways can deep learning improve operational efficiency in retail?

As you navigate the complexities of AI jargon, remember that these terms, while technical in nature, are the keys to unlocking the full potential of LLMs in your business. By demystifying these concepts, you will be better equipped to harness the capabilities of AI, driving innovation and achieving strategic objectives with confidence.

2. Go Beyond the Buzzwords for a Deeper Understanding of LLMs

In the expansive field of artificial intelligence, Large Language Models (LLMs) occupy a unique niche as a subset of Generative AI, distinguished by their remarkable ability to produce coherent and contextually aware text. Unlike traditional models that excel in specific tasks, such as object detection or fraud classification, LLMs are engineered for general-purpose language understanding, enabling them to navigate multiple domains with ease. This breadth of capability allows LLMs to generate text that spans a multitude of applications—from summarizing complex reports to translating languages with nuanced accuracy. This adaptability sets LLMs apart, as they thrive on the intricacies of human language, offering more than mere predictions or classifications. They craft narratives, engage in dialogue, and provide insights, effectively bridging the gap between human intent and machine execution.

At the heart of LLMs lies a sophisticated architecture known as the Transformer, a pivotal advancement that has redefined sequence-to-sequence tasks. Unlike its predecessors, the Transformer goes beyond the limitations of recurrent networks, employing a mechanism called self-attention to simultaneously process input tokens. This allows the model to weigh the significance of each token relative to others, like a skilled conversationalist might pay attention to the ebb and flow of dialogue, focusing on salient points while maintaining context. The importance of self-attention cannot be overstated; it is the linchpin that enables LLMs to capture dependencies and relationships within language, facilitating a more robust understanding of text. Furthermore, the efficacy of LLMs is amplified by the vast datasets on which they are trained, encompassing a diverse array of text sources that imbue the model with a rich tapestry of linguistic knowledge. These expansive datasets not only refine the model's language capabilities but also enhance its ability to generalize across various contexts and applications.

The evolution of LLMs from rudimentary sequence models to their current sophisticated architectures is a testament to the relentless pace of technological advancement. Constrained by their limited capacity to handle sequential dependencies, early models have given way to architectures capable of processing language with unprecedented depth and nuance. This transformation has been fueled by significant improvements in hardware capabilities, allowing for the training of increasingly complex models. The availability of powerful GPUs and the advent of distributed computing have democratized access to the computational resources necessary for training LLMs, enabling researchers and developers to push the boundaries of what is possible in natural language processing. These advancements have not only improved the performance and scalability of LLMs but have also opened new avenues for their application, extending their influence across industries and domains.

Despite their impressive capabilities, LLMs are often misunderstood, leading to misconceptions that can hinder their effective deployment in business contexts. One prevalent misunderstanding is the notion that LLMs are fully autonomous systems capable of replacing human decision-making entirely. While LLMs excel at generating human-like text and can automate specific tasks, they remain tools that require human oversight and guidance to ensure their outputs align with organizational goals and ethical standards. It is crucial to recognize that LLMs, like all AI systems, are susceptible to biases inherent in their training data, which can transfer into their outputs. Addressing these biases necessitates a conscientious approach to data curation and model evaluation, ensuring that the deployment of LLMs supports fair and equitable outcomes.

To fully appreciate the capabilities of LLMs, it is helpful to place them within the broader field of AI, which encompasses a spectrum of technologies ranging from rule-based automation to machine learning, deep learning, and Generative AI. Rule-based systems, the most rudimentary form of AI, operate on predefined instructions and lack the flexibility to adapt to new or unforeseen scenarios. Machine learning introduces a degree of adaptability, enabling systems to learn from data and improve over time, yet it often requires extensive labeled datasets and is typically task-specific. Deep learning, a subset of machine learning, leverages neural networks with multiple layers to process complex data representations, advancing fields such as computer vision and speech recognition. Generative AI, where LLMs reside, goes a step further by not only understanding data but also creating new content, transforming the way we interact with information and machines.

The ability of LLMs to transcend traditional AI models lies in their capacity for contextual understanding and dynamic language generation, attributes that are reshaping business communication, content creation, and strategic decision-making. As you learn more potential applications of LLMs within your organization, it is essential to approach these technologies with a nuanced understanding of their capabilities and limitations. By doing so, you can leverage LLMs to enhance operational efficiency, drive innovation, and create value, all while working within the complexities of the modern digital ecosystem.

3. Toward a Practical Vocabulary for Executives

For executives seeking to leverage AI's transformative potential, mastering this vocabulary is not just advantageous but imperative. Let's start with a fundamental term: the Application Programming Interface, or API. This concept encapsulates the protocols and tools that allow different software applications to communicate seamlessly with each other. In the context of AI, APIs are crucial for integrating AI capabilities with existing business systems. Consider a Software as a Service (SaaS) platform that leverages AI APIs for business automation; such integration can streamline operations, automate mundane tasks, and ultimately elevate productivity.

Another critical aspect involves understanding training epochs and batch sizes. These terms relate to the process of training AI models, where an epoch represents one complete pass through the entire training dataset, and batch size denotes the number of training examples utilized in one iteration. In practical terms, for a company developing a customer service AI, opting for smaller batch sizes might minimize hardware costs but could extend training time, a strategic consideration that impacts both budget allocations and project timelines.

Model accuracy and precision are metrics that gauge the performance of AI systems. In risk assessment tools, for instance, high model accuracy ensures that the AI system consistently makes correct predictions, thereby mitigating potential financial risks. On the other hand, precision measures the exactness of those predictions, ensuring that the system's outputs are relevant and applicable to specific business contexts. These metrics form the bedrock of AI system evaluation, guiding executive decisions in technology investments.

When you begin working with your first AI project and implementation terms, an "AI Use Case" becomes central to understanding how AI can address your business challenges. For instance, deploying a chatbot to automate customer support exemplifies a direct application of AI to solve a tangible problem. Before scaling such solutions, a Proof of Concept (PoC) serves as a small-scale test to determine the viability and potential impact of an AI application. This phase is crucial, providing insights into the system's capabilities and limitations and informing subsequent decisions.

Following a successful PoC, a Pilot Program offers a trial run of the AI solution in a real-world business setting before full deployment. This step is instrumental in assessing the solution's scalability—the ability of an AI system to handle increased loads or expanded operations without necessitating a complete overhaul. Effective AI integration involves seamlessly connecting AI tools with existing systems, be it Customer Relationship Management (CRM), email marketing, or inventory management software, ensuring the new technology complements rather than disrupts current workflows.

Inference, a term often encountered in AI discussions, refers to the process by which an AI model makes predictions or draws conclusions based on new data after it has been trained. This is where the actual value of AI lies—transforming raw data into actionable insights that drive business decisions.

Data forms the backbone of AI systems, and understanding data-related terms is essential for any executive. Structured Data refers to information organized in a clear format, like spreadsheets or databases, where each piece of data fits neatly into predefined categories. Customer names or sales records are typical examples. Conversely, Unstructured Data lacks this predefined structure and includes data like emails, images, and customer reviews. The latter requires sophisticated processing techniques to extract meaningful insights.

Data Labeling, the process of tagging or categorizing data, is fundamental in training AI models, as it helps the system understand and learn from the data, like marking customer inquiries as "urgent" or "general." However, data is not without its pitfalls. Data Bias occurs when training data skews results, leading to unfair or incorrect AI outputs. This problem underscores the importance of curating balanced and representative datasets. Data Privacy is another critical concern, especially in an era where data breaches and privacy laws like the General Data Protection Regulation (GDPR) and the California Consumer Privacy Act (CCPA) demand stringent compliance to protect customer and business data.

Evaluating AI performance involves several measures. Accuracy indicates how often an AI model makes correct predictions, a straightforward metric but one that speaks volumes about a system's reliability. Precision & Recall are metrics that assess the system's ability to detect relevant information without false positives. Bias & Fairness are measures that ensure AI models do not perpetuate discriminatory or skewed outcomes, a principle that aligns with ethical business practices. Explainability or Interpretability refers to how well AI decisions can be understood and justified by humans, a factor that enhances trust and transparency in AI systems.

When contemplating the financial implications of AI adoption, terms like AI-as-a-Service (AIaaS) come into play. This concept represents cloud-based AI solutions that businesses can utilize without the need to develop their own models, with examples including services from OpenAI or Google AI. The Return on Investment (ROI) is a critical measure that compares the benefits of AI implementation against the costs incurred, providing a clear picture of the financial impact. Total Cost of Ownership (TCO) goes beyond initial costs, encompassing the full spectrum of expenses associated with AI deployment, including software, data storage, and employee training.

To illustrate these terms in real-world scenarios, let's look at AI APIs in financial services for fraud detection. By integrating AI APIs, financial institutions can swiftly identify suspicious activities, safeguard assets and maintaining customer trust. Balancing training epochs and batch sizes in retail is crucial for a company seeking to enhance its product recommendation system, ensuring that AI models are cost-effective and high-performing. Healthcare is another domain where AI accuracy is paramount, particularly in diagnostic tools where precise predictions can significantly impact patient outcomes.

To stay abreast of the ever-evolving AI tools and best practices, executives should engage with recommended AI newsletters and publications, which provide insights into the latest trends and developments. Participation in AI-focused executive workshops offers opportunities for hands-on learning and networking with industry leaders. Experimenting with platforms like ChatGPT or Claude can provide practical exposure to AI technologies, enhancing familiarity and confidence in their application.

As we conclude this chapter, it is evident that mastering the language of AI empowers executives to engage in informed strategic discussions, paving the way for successful AI integration. With a firm grasp of these terms, you may now begin to navigate the complexities of AI solutions, but first let's look at the concept of resilience.

Chapter Three:
Organization Resilience With AI

"An organization's ability to learn, and translate that learning into action rapidly, is the ultimate competitive advantage."

—Jack Welch, former CEO of General Electric (GE)

Imagine a busy financial institution facilitating the flow of turbulent global markets besieged by volatility and uncertainty. Amidst this chaos, a sophisticated AI system operates in the background, quietly analyzing vast data streams, identifying patterns, and adapting to emergent trends with an agility that astonishes even the most seasoned traders. This system, fortified by the principles of resilience, has become the foundation of the institution's strategic operations, enabling it to withstand the uncertainty of market flux and continue delivering exceptional value to its stakeholders.

1. Resilience is Critical

In today's dynamic business marketplace, the resilience of AI systems is not merely advantageous but critical. As you incorporate AI into your organizational framework, it is imperative that you ensure the integrated systems can recover from failures and adapt to new data. This resilience maintains performance amidst evolving technological infrastructure. The inherent ability of AI systems to adapt and recover is part of their design, which must prioritize resilience as a core attribute to ensure consistent value delivery even in the face of adversity.

Building AI systems that emphasize resilience involves adopting design principles that enhance adaptability and scalability. A modular architecture is one such principle, facilitating the seamless updating and integration of components as technological needs evolve. Much like a well-organized assembly line incorporating new machinery without halting production, a modular AI system can integrate new data streams or algorithms without disrupting existing functionalities. Automated retraining processes enhance this flexibility and enable the system to adjust to novel data inputs, refining its models and improving its accuracy over time. By embracing these design principles, you equip your AI infrastructure with the capacity to evolve alongside your organization's strategic objectives.

Moreover, be sure to include the role of continuous monitoring and feedback loops in your pursuit of AI resilience. In the same way that a captain monitors a ship's instruments to navigate treacherous waters, real-time performance dashboards provide a window into the health of your AI systems, allowing for the timely identification and rectification of anomalies. These dashboards provide insight, offering a comprehensive view of system performance and highlighting areas that require attention or improvement. Furthermore, feedback mechanisms that incorporate user input foster an environment of continual refinement, ensuring that AI systems remain responsive to the needs and expectations of their human counterparts.

Another vital component of resilience is contingency planning, which emerges when organizations want to manage potential failures and mitigate their impact. Just as a seasoned pilot devises multiple flight paths to accommodate unexpected weather conditions, AI systems must be equipped with robust backup and failover strategies to ensure critical applications continue functioning during a system failure. These strategies involve maintaining redundant systems that seamlessly take over operations, minimizing downtime, and preserving business continuity. Scenario planning, another aspect of contingency planning, involves anticipating potential data shifts or anomalies and proactively developing strategies to address them. By envisioning various scenarios and preparing for them in advance, you bolster your organization's ability to navigate the uncertainties accompanying AI deployment.

2. Design Principles

Resilient businesses today tend to have the same five characteristics: Reliability, Security, Operational Excellence, Performance Efficiency, and Cost Optimization.

2.1   Reliability

Reliability is based on good design for business requirements, resilient design, design for recovery, and design for operations. Customers expect that their business requirements can be readily apparent, so traceability through the data flow is essential to validate compliance. A resilient design is highly scalable and available. Manage resources proactively and provide dynamic scaling capabilities to handle fluctuations in demand. Design loosely coupled modules and with good interfaces. This approach minimizes dependencies and makes the system easier to modify without introducing unanticipated faults. Build redundancy in layers to minimize single points of failure. Design in automated self-healing capabilities. These capabilities reduce the risk of faults from external factors, like human intervention. Keep it simple. Create just enough code to trace a critical path. Adhere to design and coding standards so that the system is easy for anyone to understand. Below is a table that summarizes these points. [3]

---

[3] Based on information from Google's Cloud Architecture and Microsoft's Azure Framework.

Reliability Factor	Benefit
Business Requirements	Design so that customers can easily see their requirements implemented. Allow data flow to be traced through the modules that comprise the use cases.
Provide highly scalable and available modules	Demand will fluctuate, and system modules must adapt to handle demand without human intervention.
Have loosely coupled modules with good interfaces	Reduce dependencies within the system and allow easier modifications without introducing unanticipated failures.
Build in redundancies in layers	Reduce single points of failure.
Design self-healing capabilities	Provide modules that have abilities to recover from unintended states such as unexpected human intervention.
Keep it simple	Balance the tendency to over-design and factor modules with too much granularity against having bloated modules. Adhere to design and coding standards to make modifications easier.

2.2   Security

Security involves a policy of Zero Trust and incorporates the principles of confidentiality, integrity, and availability. Creating a secure infrastructure is a continual, ongoing process. Zero trust means:

- Verify all users explicitly so that only trusted individuals are allowed to perform functions at the named location.

- Set the least access privileges for each user that are required to perform their functions.

- Play the role of an attacker to see how well your security controls perform. Design new response controls for each new risk you uncover.

Confidentiality assumes:

- Restricted access to data that is encrypted when at rest and in transit.

- Look for weak spots in authorization and authentication code.

- Block unauthorized data transfer.

- Maintain an audit log of all access activities.

Integrity means:

- Prevent data corruption by using least privilege access controls.

- Implement the standard Supply-chain Levels for Software Artifacts (SLSA) to ensure that each build has provenance and integrity.

- Trust and verify with encryption techniques like attestation, code signing, and certificates.

Availability attempts to optimize the time that the system can service the users. Some would recommend defining a Service Level Agreement (SLA). To enable this property:

- Install security protections on the recovery resources and code that are as strict as those on the primary system modules.

- Apply good design patterns and security controls on code to prevent attack vectors and software flaws.

- Implement code scanners. Always update security patches when received.

- Define Service Level Objectives (SLO) and SLA with SMART measures. They should be measurable and specific.

  o Specific - define the level of service and performance that you need.

  o Measurable - be quantifiable and trackable.

  o Achievable - be attainable with your resources and capabilities.

  o Relevant - be aligned with your business goals and objectives.

  o Time-Bound - contain a specific time frame for achievement and evaluation.

2.3 Operational Excellence

Core Principle	Benefit
Define SLOs with subsequent monitoring and testing	Achieves best availability possible
Be proactive in managing incidents and problems	Will improve code over time and help minimize downtime
Define strategies like right-sizing, autoscaling, and by using effective cost monitoring tools	Manage and optimize resources
Automate processes, streamline change management	Alleviate the burden of manual labor
Establish ongoing enhancements and introduce new solutions to stay competitive.	Demonstrate continuous improvements with innovations to stay competitive

2.4 Performance Efficiency

This principle entails a continuous monitoring and evaluation process of performance requirements and actualization. Define performance requirements for each layer of your system before you design the applications. Find elastic and scalable design patterns to help meet performance requirements. Optimize by imagining new designs or creating new features that help meet performance requirements.

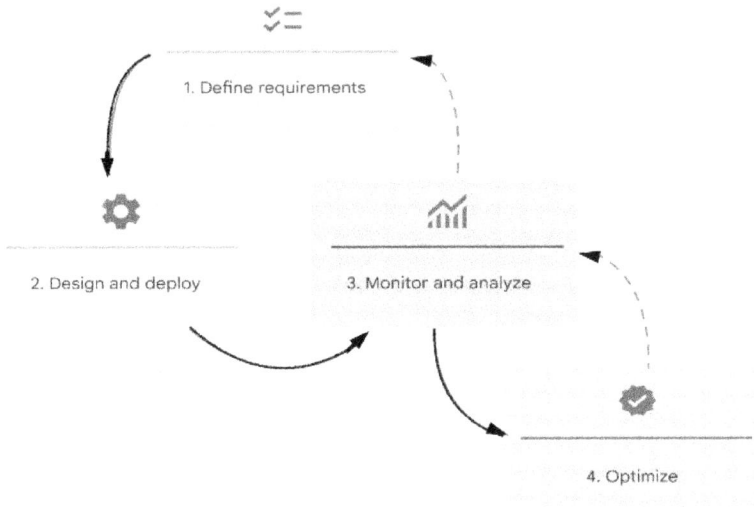

Continually monitor and analyze results with metrics, logs, tracing, and alerts.[4]

This phase should include qualitative and quantitative feedback from employees, customers, and all stakeholders. The HEART framework has been shown to reveal the level of user satisfaction with web applications and can aid in defining measures.

- Happiness is a qualitative measure that reveals a user's level of happiness with a feature. With a well-designed survey that is deployed regularly, it is possible to see changes in satisfaction over time.

- Engagement is a measure of frequency, intensity, or depth of interaction over some time period. Rather than just a count that increases indefinitely over time, an average of uses in a fixed time increment over time gives a better indication of engagement change.

- Adoption metrics monitor how many new users are found in a given time period. This measure can show

---

[4] Google, "Well-Architected Framework: Performance Optimization Pillar.

that a user has clicked into your application or you can choose to ensure that they create an account before the measure is incremented.

- Retention metrics show how many are still engaged at a later time period. This measure can be checked from week to week or with longer time periods depending on what length retention period is most meaningful.

- Task Success combines several frequently used measures such as efficiency, effectiveness, and error rate. For instance, it could be defined as time to complete, percentage of tasks completed, or number of tasks completed correctly.

2.5   Cost Optimization

Most cost optimization measures refer to the acquisition of capital expenditures or operational expenses. With the advent of cloud computing, more care needs to be taken, as working with the cloud is often merged with operational expenses. In light of this difference, follow the core principles in the following table.

Core Principle	Benefit
Align cloud computing with business objectives	Observe how business value is reflected
Ensure that employees across the organization are aware of the cost impact of their decisions	Create a culture of cost awareness throughout the business
Provide only the resources you need, and pay only for the resources that you use	Resource optimization

Monitor your usage of resources on a regular basis, and identify excess use early before it becomes a problem	Continuous optimization

Interactive Element: Resilience Checkpoint

To assess the resilience of your current AI systems, consider the following questions as a checkpoint:

1. Is your AI infrastructure designed with modular components that facilitate easy updates and integration?

2. Do you have automated retraining processes in place to adapt to new data inputs?

3. Are real-time performance dashboards actively utilized to monitor system health and identify anomalies?

4. Have you implemented feedback mechanisms that incorporate user input for continual system improvement?

5. Do you have robust backup and failover strategies to ensure business continuity during system failures?

6. Is scenario planning part of your contingency strategy to address potential data shifts or anomalies?

Reflecting on these questions will provide valuable insights into the resilience of your AI systems and highlight areas for enhancement.

By prioritizing resilience in designing and implementing AI systems, you lay the foundation for sustained organizational growth and innovation. Resilient AI systems withstand disruptions and thrive amidst them, adapting to changing needs and delivering consistent value. As you continue to integrate AI into your business operations, embracing these resilience principles will empower your organization to navigate the challenges of the digital age with confidence and agility, ensuring that your strategic objectives are met and exceeded.

3. Cultural Shifts: Embracing Change in the AI Era

Imagine a traditional company where the decision-making process is firmly entrenched in a hierarchical structure, with directives cascading from the top down. While effective in a more stable business environment, such a configuration often proves inadequate in the face of rapid technological change, particularly when integrating artificial intelligence. AI requires an organizational culture that is flexible and open to innovation and experimentation. This cultural shift necessitates a move from rigid hierarchies to more agile frameworks, where decision-making is decentralized, and teams have the autonomy to explore new ideas and iterate rapidly. Agile structures empower employees at all levels to contribute to AI initiatives, fostering an environment where creativity and innovation can flourish without the constraints of traditional top-down management.

Organizations must encourage a culture where employees feel safe to take calculated risks and learn from failures without fear of retribution. This shift towards a more experimental approach mirrors the iterative nature of AI development itself, where models are continually refined based on feedback and new data. By embedding this mindset into the organizational fabric, you enable your teams to adapt quickly to technological advancements and seize new opportunities as they arise. The transformation is not merely about adopting new technologies but about reshaping the very character of the organization to embrace change as a catalyst for growth.

When fostering a culture receptive to change, a business must implement targeted change management programs focused on AI. These programs should address the technical aspects of AI integration and the human elements, preparing employees to embrace new workflows and processes. Training sessions and workshops can be instrumental in helping employees understand the impact of AI on their roles and the organization as a whole. Furthermore, creating innovation hubs or labs within the organization can serve as incubators for new ideas and technologies, providing a dedicated space for experimentation and collaboration. These hubs encourage cross-functional teams to work together, breaking down silos and fostering a culture of shared learning and innovation.

Communication is pivotal in facilitating cultural shifts and ensuring alignment across the organization. Regular town halls and workshops provide a platform for leaders to discuss the implications of AI, addressing both opportunities and challenges. These forums also serve as an opportunity to celebrate successes, share lessons learned, and reinforce the organization's commitment to AI-driven transformation. In addition, open forums for employees to voice their concerns and suggestions can help alleviate apprehensions and build trust, creating an environment where everyone feels valued and heard. Transparent communication fosters a sense of collective ownership and accountability, ensuring the entire organization aligns in its pursuit of AI excellence.

Integrating AI literacy into company culture is another critical component of driving cultural change. AI literacy empowers employees by demystifying the technology and enabling them to engage meaningfully with AI initiatives. Providing AI training modules accessible to all employees, regardless of their technical background, is essential for fostering a culture of inclusivity and collaboration. These modules should cover foundational AI concepts and practical applications, equipping employees with the knowledge to contribute to AI projects and initiatives. Furthermore, incorporating AI topics into regular team meetings and discussions reinforces the importance of AI in the organization's strategic objectives and keeps AI at the forefront of employees' minds.

As you navigate the integration of AI into your organizational culture, consider the broader implications of these cultural shifts. Embracing AI is not solely about adopting new technologies; it is about transforming how your organization thinks, operates, and interacts with its environment. This transformation requires a commitment to change and a willingness to challenge established norms and embrace new paradigms. By fostering a culture that values flexibility, innovation, and collaboration, you position your organization to thrive in the AI era, ready to harness the full potential of these transformative technologies.

4. Strategies for Resilient Leadership in AI

Resilient leadership is a cornerstone of an agile organization that is part of AI adoption. Certain qualities become indispensable when you steer your organization through this transformative phase. Emotional intelligence, for instance, stands at the forefront, offering leaders the ability to navigate the nuanced interpersonal dynamics that AI integration inevitably affects. This quality involves understanding and managing one's emotions and those of others to foster a collaborative environment where team members feel valued and understood. Alongside this, adaptability in decision-making is crucial, as leaders must be prepared to pivot and recalibrate strategies in response to the rapid technological shifts and unforeseen obstacles characteristic of AI initiatives. Visionary thinking, when combined with practical execution skills, allows leaders to articulate a compelling future while grounding their strategies in actionable, realistic steps. This duality—seeing the big picture while managing the minutiae— sets apart leaders who can inspire team resilience and drive successful AI transformations.

Leadership development programs tailored to the AI domain are instrumental in equipping leaders with the skills needed to manage the intricacies of AI-related changes effectively. Workshops focusing on AI ethics and governance provide a platform for leaders to engage with the ethical considerations and regulatory frameworks that are increasingly important in AI deployment. These sessions highlight the moral imperatives of AI usage and offer guidance on implementing

AI that aligns with organizational values and societal norms. Additionally, mentorship programs that pair seasoned leaders with AI novices are helpful for knowledge exchange and growth. Through these relationships, emerging leaders gain insights into AI integration's strategic and operational aspects, benefiting from the experiences and wisdom of those who have successfully navigated similar challenges. Such programs foster a culture of continuous learning and development, ensuring that your leadership team is well-equipped to guide the organization through the AI era.

Diversity within leadership teams is essential to foster innovation and resilience in AI initiatives. By encouraging diverse perspectives in AI strategy discussions, leaders can draw from a rich tapestry of experiences and viewpoints, which enhances problem-solving and creativity. This diversity is not limited to demographic markers but extends to cognitive and experiential differences, creating a dynamic environment where varied ideas can flourish. Fostering inclusivity in AI project leadership roles further amplifies this effect, ensuring that all voices are heard and valued. An inclusive leadership team can anticipate and address the multifaceted challenges of AI integration, from ethical considerations to technical hurdles. By embracing diversity, you cultivate a resilient leadership framework adaptable to the complexities of AI-driven transformation.

Consider the case of a tech company CEO who successfully spearheaded a major AI-driven pivot within their organization. Faced with a declining market share due to increased competition, the CEO recognized the potential of AI to enhance their product offerings and streamline operations. So, the CEO navigated the organization through a comprehensive AI integration process by assembling a diverse leadership team and leveraging their collective expertise. The AI initiative included deploying machine learning algorithms to optimize supply chain management and incorporating AI-driven analytics into customer relationship management systems. The result significantly improved operational efficiency and customer satisfaction, positioning the company as a formidable player in the tech industry. This case exemplifies how resilient leadership, characterized by diversity and strategic foresight, can drive successful AI transformations.

In the healthcare sector, a visionary leader implemented AI technologies to enhance patient care outcomes. By integrating AI into diagnostic processes, the leader was able to reduce diagnostic errors and improve treatment accuracy. AI systems capable of analyzing medical images and patient data with unprecedented precision achieved this success. The leader's commitment to promoting a culture of innovation and inclusivity was instrumental in overcoming initial resistance to AI adoption among healthcare professionals. By involving diverse stakeholders in the decision-making process and providing ongoing training and support, the leader ensured a smooth transition to AI-enhanced practices. This case illustrates the potential of resilient leadership to harness AI for transformative impact in healthcare, improving both patient outcomes and operational efficiencies.

As we explore these strategies for resilient leadership in AI, it becomes evident that the qualities and skills required to affect the AI transformation extend beyond traditional leadership paradigms. Emotional intelligence, adaptability, visionary thinking, and diversity are not merely desirable traits but essential components of a leadership framework that can thrive amidst the complexities of AI integration. By investing in leadership development programs and encouraging a culture of inclusivity and continuous learning, you empower your leadership team to navigate the challenges of the AI era with confidence and agility.

At this writing, we offer suggestions for leading online or location-based leadership courses:

- Oxford Management Centre

- MIT Management Executive Education

- Stanford Business Executive Education

- CalTech Center for Technology and Management

In this chapter, we've examined the qualities and strategies that define resilient leadership for AI adoption. Leaders can effectively guide their organizations through AI-driven transformations by cultivating emotional intelligence, adaptability, and visionary thinking and emphasizing diversity and inclusive practices. As we move forward, the focus shifts to exploring the practical applications and strategies enabling businesses to integrate AI systems successfully, enhancing operational efficiency and driving innovation across various sectors.

Chapter Four:
Engaging Stakeholders and Building Support

"The purpose of a company is to engage all its stakeholders in shared and sustained value creation. In creating such value, a company serves not only its shareholders, but all its stakeholders – employees, customers, suppliers, local communities and society at large. The best way to understand and harmonize the divergent interests of all stakeholders is through a shared commitment to policies and decisions that strengthen the long-term prosperity of a company."[5]

—Davos Manifesto 2020

In the boardroom of a top-500 company, diverse voices congregate to shape the future of a cutting-edge AI initiative. The energy is palpable as executives, department heads, and project managers gather to discuss the transformative potential of Large Language Models (LLMs) within their organization. Yet, amidst the excitement, a common challenge emerges—the imperative of securing buy-in and consensus from internal and external stakeholders. Achieving agreement from all stakeholders is no small feat, for the path to successful AI implementation is paved with the complexities of aligning varied interests and perspectives. As a leader, you understand that stakeholder engagement is not merely a box to check but a strategic endeavor, crucial to the enduring success of your AI projects.

---

[5] World Economic Forum, Davos Manifesto 2020.

1. Stakeholder Roles

In the intricate process of AI integration, identifying key stakeholders becomes your first step for turning your vision into reality. Internally, this group includes executives who drive strategic direction, department heads who oversee operational execution, and project managers who orchestrate the myriad tasks that bring your AI vision to life. Each of these roles plays a pivotal part, and their support is paramount to fulfilling the multifaceted requirements of AI. Externally, your stakeholders extend to customers who will ultimately benefit from AI-driven enhancements, partners whose collaboration can amplify your efforts, and regulators who ensure compliance with evolving legal standards. Engaging these diverse groups requires a nuanced understanding of their unique perspectives and concerns, as their collective input is invaluable in shaping AI solutions that are not only effective but also equitable and sustainable.

Developing a stakeholder engagement plan is your blueprint for ensuring alignment and support throughout the AI project lifecycle. Regular stakeholder meetings and workshops are the backbone of this plan, providing a forum for open dialogue, knowledge exchange, and collaborative problem-solving. These gatherings are not mere formalities but opportunities to build trust, foster transparency, and cultivate a shared sense of ownership over the project's outcomes. Stakeholder mapping is essential in this endeavor, offering a visual representation of the various interests and influences at play. By understanding who holds sway over key decisions and who will be impacted by the AI initiative, you can tailor your engagement strategies to address their specific needs and priorities. This strategic foresight enables you to anticipate and address potential roadblocks proactively, smoothing the path to successful implementation.

There have been many different suggestions for defining stakeholder mappings, but one stands out as a practical tool for establishing meaningful relationships. Mitchell, Agle, and Wood describe a "Salience" model as a way of understanding "who and what really counts," in Toward a Theory of

Salience Model for Stakeholder Identification

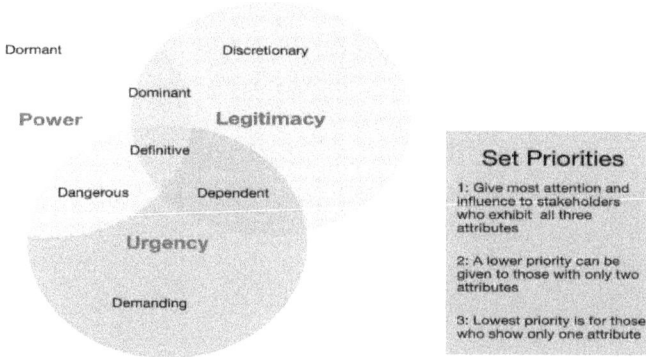

Dormant

Discretionary

Dominant

Power

Legitimacy

Definitive

Dangerous

Dependent

Urgency

Demanding

Set Priorities

1: Give most attention and influence to stakeholders who exhibit all three attributes

2: A lower priority can be given to those with only two attributes

3: Lowest priority is for those who show only one attribute

Stakeholder Identification and Salience: Defining the Principle of Who and What Really Counts. Stakeholder attributes are often shown in a Venn diagram, where the power to influence decisions, legitimacy of interest, and urgency of need can overlap to give more insight. Salience is the quality where a stakeholder exhibits multiple attributes. A stakeholder with Power can impose their will upon decisions, however we see that power can be Dominant together with Legitimacy or can wane and be Dormant when it is the only attribute. Legitimacy usually refers to a contractual interest or a user internal to the organization. When Power and Legitimacy overlap into Dominant, another way to think of it is authoritative. Without a contractual interest, the relationship can be Discretionary. Urgency reflects a stakeholder who has a real and critical need within a specified time frame. When a legitimate stakeholder has an urgent and critical need, then they are Dependent on the coalition's results. When a power relationship exists, then it can also be Dangerous to the existence of the coalition when it is urgent. An Urgent relationship is always Demanding. Definitive stakeholders are the core managers who are leading an initiative. They hold relationships with all the stakeholders and make decisions about priorities.

Establishing a roster of stakeholders should be an early task of a project manager. This roster documents all the attributes for each stakeholder such as name, relationship to the project, primary interest, and priority. It acts as a reference document when later decisions must be made in workshops.

Encouraging collaborative decision-making is crucial to harnessing the collective wisdom of your stakeholders. Joint workshops for creating AI strategies invite stakeholders to contribute their insights and expertise, fostering a sense of agency and investment in the project's success. These workshops are a means of generating innovative ideas and a platform for building consensus and resolving conflicts. Cross-functional task forces for project oversight further enhance collaboration, bringing together diverse perspectives to ensure that the AI initiative remains aligned with organizational goals and adapts to changing circumstances. By involving stakeholders at every project stage, you create a culture of inclusivity and empowerment, where all voices are heard and all contributions are valued.

Addressing stakeholder concerns and feedback is critical to effective engagement, as it demonstrates your commitment to responsive and adaptive leadership. Feedback loops with structured response protocols allow you to capture and analyze stakeholder input, ensuring that their concerns are addressed promptly and integrated into the AI initiative. This iterative feedback and refinement process not only enhances the quality and relevance of the AI solutions but also builds trust and confidence among stakeholders. Surveys and polls are valuable tools for gauging stakeholder sentiment, providing quantitative data that can inform your engagement strategies and highlight areas for improvement. By maintaining an open line of communication and demonstrating a willingness to adapt based on stakeholder feedback, you create a dynamic and resilient AI project that is poised for success.

Interactive Element: Stakeholder Mapping Exercise

To effectively engage your stakeholders, consider conducting a stakeholder mapping exercise to visualize the various influences and interests within your AI project. This exercise will help you identify key stakeholders, assess their levels of influence and interest, and develop targeted engagement strategies. Use the following steps to guide your mapping process:

1. Identify Stakeholders: List all potential stakeholders, both internal and external, who may be impacted by or have an interest in your AI initiative.

2. Assess Influence: Evaluate each stakeholder's level of influence over the project's outcomes, considering factors such as decision-making authority, expertise, and resources.

3. Evaluate Interest: Determine each stakeholder's level of interest in the AI initiative, considering their motivations, concerns, and potential benefits.

4. Prioritize Engagement: Based on your assessments, prioritize stakeholders for engagement, focusing your efforts on those with high influence and interest.

5. Develop Strategies: Tailor your engagement strategies to address each stakeholder's specific needs and priorities, ensuring that their voices are heard and their contributions are valued.

By completing this exercise, you will better understand the stakeholder roles and influences and be better equipped to navigate the complexities of AI integration with confidence and agility. As you engage with your stakeholders, remember that their support is not merely a means to an end but a vital component of the AI initiative's success. Through thoughtful and inclusive engagement, you can harness the collective power of your stakeholders to drive innovation, mitigate risks, and ensure that your AI solutions are meaningful and impactful for all involved.

2. Communication Strategies: Articulating AI Benefits

Communicating the intricacies of AI, particularly Large Language Models (LLMs), demands a nuanced and multifaceted approach. You must tailor your communication strategies to resonate with the diverse audiences that comprise your stakeholder ecosystem. Executive briefings should focus on the strategic benefits of AI, emphasizing how LLMs can drive competitive advantage, enhance operational efficiency, and open new revenue streams. These discussions should articulate the alignment of AI initiatives with broader organizational goals, using clear, concise language that underscores the potential for ROI and long-term growth. Meanwhile, technical sessions for IT teams should go into the granular details of implementation plans, providing the technical insight necessary for seamless integration. These sessions must address system architecture, data requirements, and integration protocols, ensuring that your technical staff is fully equipped to execute the AI strategy effectively.

The art of storytelling serves as a powerful tool in conveying the transformative impact of AI, making complex concepts relatable and engaging. By weaving narratives that illustrate the real-world implications of AI, you can captivate your audience and foster a deeper understanding of its potential. Share success stories of AI-driven transformations, highlighting tangible outcomes such as increased productivity, enhanced customer satisfaction, or improved decision-making capabilities. These stories should be supported by quantitative data that underscores the effectiveness of AI solutions, lending credibility and weight to your narrative. Personal narratives from employees who have benefited from AI can further humanize the technology, illustrating how it empowers individuals to focus on higher-value tasks and make more informed decisions. These personal accounts can resonate deeply with your audience, bridging the gap between abstract technology and its practical, everyday applications.

Visual aids and demonstrations can significantly enhance your communication strategy, providing stakeholders with a dynamic and immersive understanding of AI capabilities. Interactive demos of AI applications in action allow stakeholders to experience firsthand the functionalities and benefits of LLMs, fostering a sense of engagement and curiosity. Through these demonstrations, stakeholders can witness the efficiency of AI in real-time, whether it's automating routine tasks, generating insights from data, or facilitating seamless customer interactions. Infographics are another powerful visual tool that distills complex AI performance metrics into accessible, digestible formats. These visuals can highlight key performance indicators, such as accuracy, speed, and cost savings, offering stakeholders a clear snapshot of AI's impact on business operations. Visual aids not only enrich the communication process but also aid in retaining stakeholder interest and understanding.

Maintaining open communication channels is vital for keeping stakeholders informed about the progress of AI initiatives. Regular updates on project milestones, challenges, and successes help build trust and transparency, ensuring stakeholders remain engaged and supportive throughout the AI project lifecycle. These updates can be disseminated through various channels, such as newsletters, webinars, or collaborative platforms, allowing stakeholders to access information that suits their preferences. Providing stakeholders with timely updates also offers an opportunity to address any emerging concerns or feedback, fostering a culture of collaboration and continuous improvement. Committing to transparent and proactive communication can nurture a robust stakeholder relationship, paving the way for a successful and impactful AI integration.

3. Overcoming Resistance: Fostering a Pro-AI Culture

It is not uncommon to encounter a degree of apprehension when introducing AI technologies into an organization. This hesitance often stems from a fear deeply rooted in the human psyche: the potential for job displacement. Employees frequently perceive AI as a formidable force poised to render their roles obsolete, a faceless entity encroaching upon their professional territories. Engaging with this fear requires a nuanced approach that acknowledges the validity of these concerns while offering reassurance through a carefully crafted narrative. It is imperative to communicate that AI, rather than replacing human roles, serves as an augmentative tool that alleviates menial tasks and allows employees to channel their energies into more strategic and creative endeavors. By fostering an environment where employees see AI as a collaborative partner rather than a competitor, you can transform anxiety into acceptance, empowering your workforce to embrace technological advancements enthusiastically rather than fearfully.

Another prevalent source of resistance is skepticism regarding AI's reliability and accuracy. This skepticism is not unfounded; AI systems, much like any other technology, are susceptible to errors and biases. These imperfections often reflect the data on which they are trained and the algorithms that drive their operations. Addressing this skepticism necessitates a transparent dialogue about AI's limitations and potential pitfalls, coupled with a commitment to continuous improvement and ethical considerations. You can foster a culture of trust and confidence in AI's capabilities by openly discussing these challenges and outlining the measures to mitigate them—such as rigorous testing, bias reduction efforts, and ongoing system refinement. Moreover, showcasing data-driven successes and tangible outcomes achieved through AI initiatives can further reinforce the technology's credibility, transforming skepticism into advocacy.

Addressing these sources of resistance head-on by adopting targeted strategies is essential to encourage a pro-AI culture. Education plays a pivotal role in this effort, serving as the cornerstone of AI literacy within the organization. By offering workshops, seminars, and interactive training sessions, you can equip employees with the knowledge and skills they need to work through AI challenges with confidence. These educational initiatives should cover the technical aspects of AI and explore its ethical implications and potential societal impact, providing a holistic understanding of the technology. By demystifying AI and fostering a sense of curiosity and engagement, you can transform resistance into a willingness to learn and adapt.

Involvement is another key strategy for overcoming resistance, as it invites employees to participate actively in AI initiatives and decision-making processes. Involving team members in developing and implementing AI projects instills a sense of ownership and agency, empowering them to contribute their insights and expertise. This participatory approach enhances the quality and relevance of AI solutions and fosters a collaborative culture where diverse perspectives are valued and respected. Encouraging cross-functional collaboration and creating opportunities for employees to engage with AI projects firsthand can further solidify their commitment to the technology's success.

Trust and transparency are foundational elements in fostering a pro-AI culture, as they create an environment where employees feel secure and supported. Establishing trust requires consistent and transparent communication about AI initiatives, including their objectives, challenges, and progress. Maintaining an open line of communication and providing regular updates can alleviate concerns and build confidence in the organization's AI strategy. Additionally, cultivating a culture of transparency involves acknowledging and addressing any setbacks or failures and demonstrating a commitment to accountability and continuous improvement.

Engagement is the final pillar in overcoming resistance to AI adoption, as it involves creating opportunities for ongoing dialogue and feedback. Forums, Q&A sessions, and informal discussions allow employees to voice their concerns, ask questions, and share their experiences with AI. These engagements facilitate the exchange of ideas and insights and reinforce the organization's commitment to inclusivity and collaboration. By actively listening to employee feedback and incorporating their input into AI initiatives, you can foster a culture of trust and mutual respect, paving the way for successful AI integration.

As we conclude this chapter, it is evident that overcoming resistance to AI adoption requires a multifaceted approach, one that addresses the fears, skepticism, and concerns of stakeholders with empathy and understanding. By prioritizing education, involvement, trust, and engagement, you can foster a pro-AI culture that embraces technological advancements with enthusiasm and resilience. This cultural shift is not only essential for the successful integration of AI but also for positioning your organization at the forefront of innovation and progress. As we transition to the next chapter, we will explore the strategic implementation of LLMs and how to integrate these advanced technologies into your business processes effectively.

Chapter Five:
Measuring Success: Key Metrics for LLM ROI

"You can't manage what you can't measure."

—Peter Drucker

"The goal is to turn data into information, and information into insight."

—Carly Fiorina, former CEO of Hewlett Packard

Imagine yourself in the conference room of a thriving financial firm. The executives have gathered to review the latest quarterly performance report. At the center of the conversation is the firm's recent investment in Large Language Models (LLMs), and the question on everyone's mind is whether this cutting-edge technology is delivering tangible returns. The firm's CEO stands up, ready to present a set of meticulously curated metrics that will shed light on the AI's performance, offering not just numbers but a narrative of innovation and transformation. This is the power of clear, insightful metrics: they turn abstract investments into concrete results, providing a roadmap for strategic decisions and future growth.

1. Key Measures

In the world of AI, where complexity often obscures clarity, it is crucial to distill performance into understandable metrics that capture the essence of what these systems achieve. Accuracy, a fundamental metric, is the linchpin of AI performance, offering a straightforward measure of how frequently an AI model makes correct predictions. It encapsulates the model's overall reliability, guiding

53

stakeholders in assessing its value across various applications. For an LLM deployed in customer service, for instance, high Accuracy translates into correct responses to customer queries, enhancing satisfaction and efficiency. It is often expressed as a percentage.

Accuracy = Number of Correct Predictions/Total Number Predictions * 100

Yet, Accuracy alone cannot capture the nuances of an AI's capability—Precision steps in to refine this understanding, measuring the proportion of true positives among all possible positive predictions (true positives + false positives). In practical terms, consider an email filtering system: precision ensures that flagged spam emails are indeed spam, minimizing false alarms that could disrupt workflow. This specificity is crucial in environments where the cost of false positives is high, such as in financial fraud detection or medical diagnosis, where Precision ensures that only genuine threats or conditions are highlighted for action.

Precision = Number of True Positives/Total Number of Positives

Complementing Precision is Recall, or Sensitivity, which assesses the system's ability to identify all relevant cases within a dataset. Recall is paramount in scenarios where missing a positive instance could have serious repercussions—such as detecting fraudulent transactions or rare medical conditions. By focusing on capturing all potential positives, Recall ensures that the AI system does not overlook critical events, even at the risk of including some false positives.

Recall = True Positives / (True Positives + False Negatives)

Balancing Precision and Recall is the F1 Score, a harmonic mean offering a single metric when false positives and false negatives carry significant costs. This balanced approach is beneficial in domains where accuracy and coverage are equally important, providing a holistic view of the model's efficacy.

The F1 Score becomes important when distilling complex performance dynamics into a comprehensible figure that aids in strategic decision-making.

$$F1 = 2 * (Precision * Recall) / (Precision + Recall)$$

Another important classification metric is the ROC-AUC (Receiver Operating Characteristic - Area Under Curve), where the name is a holdover from WWII radar detection. This measure evaluates a model's ability to distinguish between classes. A higher AUC indicates strong performance in differentiating between categories, an invaluable feature in applications like credit scoring or risk assessment, where distinguishing between risk levels is critical. This metric provides a visual and quantitative measure of the model's discriminative power, offering insights that support nuanced evaluations of AI performance.

TPR = Correctly Classified Actual Positives/Total Actual Positives

FPR = Incorrectly Classified Actual Negatives/Total Actual Negatives

Graphing TPR against FPR at different thresholds can reveal the best thresholds for model predictions.If false positives (false alarms) are highly costly, it may make sense to choose a threshold that gives a lower FPR, like the one at point A, even if TPR is reduced. Conversely, if false positives are cheap and false negatives (missed true positives) highly costly, the threshold for point C, which maximizes TPR, may be preferable. If the costs are roughly equivalent, point B may offer the best balance between TPR and FPR.

Shifting from classification to regression metrics, we encounter tools designed to assess predictive accuracy in forecasting and continuous output domains. Mean Absolute Error (MAE) offers a straightforward measure of the average difference between predicted and actual values, clearly indicating model precision. This metric may mean more to you as a measure of the average error magnitude. A low value of MAE means that the model is making accurate predictions.

$$MAE = 1/N\sum |actualvalue - predictedvalue|$$

where N is the total number of values.

Mean Squared Error (MSE) penalizes larger discrepancies more heavily by squaring the errors, thereby highlighting areas where the model may falter in capturing outlier events. These metrics are vital in sectors like finance and supply chain management, where predictive accuracy directly influences strategic planning and resource allocation.

$$MSE = 1/N\sum (actualvalue - predictedvalue)^2$$

where N is the total number of values.

The difference between the MAE and MSE formulas is simply in the squaring of the value differences. Squaring the differences will shrink small numbers and enlarge bigger numbers. So, in choosing to use one formula over the other, consider whether your data set contains some large outliers. If so, then you should choose MAE. [6]

The R^2 Score, or R-squared, gauges how well the model explains variability within the data, with values closer to one indicating superior explanatory power. This metric is particularly relevant in contexts where understanding the underlying data dynamics is essential, such as economic forecasting or market analysis, where a high R^2 Score translates into actionable insights and informed decision-making.

First calculate the sum of the squares of the difference between the actual values and the predicted values (SSres).

---

[6] Google for Developers, "Machine Learning Crash Course: Linear Regression."

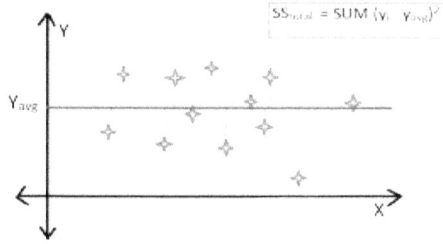

Next calculate the sum of the squares of the difference between the actual values and the mean of the predicted values (SSTotal).

Then R^2 can be calculated as:

$$R^2 = 1 - \frac{SS_{RES}}{SS_{TOT}} = 1 - \frac{\sum_i (y_i - \hat{y}_i)^2}{\sum_i (y_i - \bar{y})^2}$$

Be aware that this metric is useful but not sufficient to understand how well the model has predictive power. This metric is prone to overfitting, where it can explain almost all of the training data but not new data. Use this metric in conjunction with MAE or other regression metrics to prevent making false conclusions. [7]

Efficiency metrics, meanwhile, assess how swiftly and effectively an AI system processes data, a critical factor in operational contexts where speed is of the essence. Consider creating a set of use cases to use as a benchmark, then clock the time to complete for each of them.

---

[7] Orkun Orulluoğlu, "R-Square in Machine Learning: A Powerful Tool for Evaluating Model Performance."

Inference time, which measures the duration between input and output generation, is crucial for applications like chatbots, where rapid response times enhance user experience and satisfaction. Throughput, reflecting the number of tasks an AI can handle per second, becomes a key consideration in high-volume environments like customer service centers or real-time analytics, where capacity and speed dictate success.

Latency, the delay between data input and system response, is another critical metric, especially for real-time applications such as autonomous vehicles or live customer interactions, where delays could have significant implications.

Scalability, the system's ability to maintain performance under increasing workloads, becomes vital as organizations expand their AI applications. A scalable model ensures that as data volumes grow, the system can adapt without sacrificing efficiency or accuracy, providing a sustainable foundation for long-term AI integration.

As you consider these metrics, it becomes clear that they are more than mere numbers; they are the language through which AI communicates its value to the business. Effective metric selection and interpretation turn data into insights, guiding strategic decisions and ensuring that AI investments yield tangible returns. To aid in this process, I invite you to engage with the following exercise to prompt reflection on your organization's AI metrics strategy.

Reflection Exercise: Assessing Your AI Metrics Strategy

1. **Identify Key Metrics:** What primary metrics do you currently use to evaluate your LLM's performance? Consider their relevance and effectiveness in capturing the AI's value.

2. **Align with Objectives:** Do these metrics align with your strategic business objectives? Reflect on how they support or hinder your organization's goals.

3. **Evaluate Balance:** Is there a balance between classification and regression metrics? Consider whether your current approach provides a comprehensive view of AI performance.

4. **Consider Efficiency:** How are efficiency metrics integrated into your evaluation process? Reflect on their role in optimizing operational performance and user experience.

5. **Revisit Scalability:** How does scalability feature in your metrics strategy? Consider its importance as your AI applications evolve and expand.

Through this exercise, you can better understand your organization's metrics strategy, identify areas for enhancement, and ensure that your AI initiatives are prepared for success.

2. Crafting a Convincing Business Case for LLM Investment

Imagine walking into a boardroom filled with anticipation, where the future of your organization hinges on a decision to invest in Large Language Models (LLMs). As an executive, you are tasked with presenting a compelling business case that not only elucidates the strategic value of this investment but also addresses the nuanced intricacies of implementation and risk management. To win approval, you must structure your business case meticulously, offering a narrative that resonates with stakeholders and aligns with the broader organizational vision.

Begin your business case with an executive summary that captures the strategic importance of LLMs in your organization. This section should succinctly articulate the overarching goals of the investment, providing a high-level overview of how LLMs can drive innovation, enhance efficiency, and bolster competitive advantage. The executive summary serves as a roadmap for what follows, setting the stage for a deeper exploration of the benefits and implications of LLM integration. By clearly defining the strategic intent, you provide stakeholders with a lens to view the following detailed analyses, framing the discussion regarding long-term value and alignment with corporate objectives.

The core of your business case should offer a detailed analysis of costs versus projected benefits, meticulously breaking down the financial implications of LLM implementation. This analysis requires a comprehensive examination of direct and indirect costs, including software acquisition, infrastructure upgrades, and training expenses. However, it is equally important to account for the anticipated benefits, such as operational efficiencies, enhanced customer engagement, and potential revenue streams. Quantifying these benefits provides a clear picture of the return on investment, enabling stakeholders to make informed decisions based on empirical data and projected outcomes. It's essential to present these figures transparently and precisely, ensuring stakeholders understand both the short-term financial commitments and the long-term gains.

Recognizing the inherent risks of LLM investments is crucial to crafting a credible business case. Identify potential technological risks, such as integration challenges, data privacy concerns, and the possibility of algorithmic bias. Each risk should be accompanied by a well-considered mitigation strategy, demonstrating your proactive approach to managing uncertainties. For instance, strategies include establishing robust data governance frameworks, implementing continuous monitoring and evaluation protocols, and investing in employee training to mitigate skill gaps. By addressing risks head-on, you instill confidence in stakeholders, assuring them that you are prepared to navigate the complexities of LLM deployment while safeguarding organizational interests.

A compelling business case also emphasizes strategic alignment with the organization's goals, illustrating how LLM investments support broader digital transformation initiatives and contribute to competitive positioning. Highlight the harmonies between LLM capabilities and existing business strategies, showcasing how AI-driven insights can drive decision-making, optimize processes, and enhance customer experiences. For example, if your organization is committed to digital transformation, demonstrate how LLMs can accelerate this journey by automating routine tasks, providing real-time data analysis, and enabling personalized customer interactions. By articulating these connections, you reinforce that LLM investments are not isolated endeavors but integral components of a cohesive strategic vision.

To substantiate your business case, present real-world examples and success stories that validate the effectiveness of LLMs in similar contexts. Case studies provide tangible evidence of LLM benefits, offering insights into the transformative impact of AI on organizational performance. For instance, consider a case study of a retail company that leveraged LLMs to enhance customer engagement, resulting in increased sales and improved customer satisfaction. Highlighting such examples demonstrates the practical applicability of LLMs and reassures stakeholders of their value. Similarly, share the success story of a manufacturing firm that achieved significant cost reductions in supply chain management through LLM-driven process optimization. By showcasing these successes, you provide a compelling narrative that complements the quantitative analyses, appealing to rational and emotional aspects of stakeholder decision-making.

In constructing your business case, recognize the importance of tailoring your narrative to resonate with diverse stakeholder perspectives. Executives may be most interested in strategic alignment and financial returns, while IT teams may focus on technical feasibility and risk mitigation. By anticipating and addressing these varied interests within your business case, you ensure that all stakeholders feel heard and engaged. This holistic approach fosters a sense of shared ownership, increasing the likelihood of consensus and support for the proposed investment.

Ultimately, a well-crafted business case for LLM investment is a powerful tool for guiding organizational decision-making, providing both a vision for the future and a roadmap for achieving it. Through careful analysis, strategic alignment, and compelling storytelling, you can effectively communicate the transformative potential of LLMs, positioning them as catalysts for innovation and growth within your organization. As you embark on this journey, remember that the strength of your business case lies in its ability to bridge the gap between technical complexity and strategic clarity, offering stakeholders a comprehensive understanding of the value and impact of LLM integration.

3. Communicating Value: ROI Presentations to Stakeholders

Communicating the value of Large Language Models (LLMs) demands a strategic approach to communication that resonates with diverse stakeholder groups. Tailoring ROI presentations to these groups is not merely an exercise in persuasion; it is an opportunity to align the multifaceted capabilities of LLMs with your audience's varied interests and concerns. For executive leadership, the presentation must transcend technical jargon, focusing instead on strategic priorities that resonate with their overarching vision for the organization. This involves crafting a narrative highlighting how LLMs contribute to competitive advantage and long-term growth, emphasizing key outcomes such as market positioning and operational efficiencies. By contextualizing the ROI within the broader set of strategic goals, you enable executives to grasp LLMs' tangible and intangible benefits, fostering informed decision-making that aligns with organizational objectives.

Conversely, when addressing technical teams, the focus shifts toward the granular details of implementation and functionality. Technical teams are often concerned with the practicalities of integration, system performance, and potential challenges. Here, the presentation should go into how LLMs interact with existing infrastructure, detailing the technical innovations that underpin their efficacy. Providing insights into LLMs' scalability, adaptability, and security can assuage concerns and build confidence among technical stakeholders, ensuring that they are equipped to support the implementation process. By customizing your communication strategy to address each stakeholder group's unique perspectives and priorities, you create a cohesive narrative that underscores the value of LLMs, engendering trust and buy-in across the organization.

The power of visual aids and data visualization in enhancing the understanding of ROI outcomes cannot be overstated. In today's data-driven business environment, visuals such as infographics and dashboards offer a compelling means of conveying complex information in an accessible format. Infographics can succinctly illustrate cost savings, efficiency gains, and other key performance indicators, visually representing the quantitative benefits of LLM investments. These visuals are a focal point for discussions, anchoring the narrative in empirical evidence and facilitating a shared understanding among stakeholders. On the other hand, dashboards provide real-time performance metrics, offering a dynamic tool for tracking the ongoing impact of LLMs on business operations. Integrating these visual elements into your presentations transforms abstract data into actionable insights, enabling stakeholders to engage with the material meaningfully.

Storytelling emerges as a transformative tool in ROI presentations, bridging the gap between data and human experience. By weaving narratives that showcase before-and-after scenarios, you can vividly demonstrate the transformative impact of LLMs within the organization.

Consider crafting a story highlighting a specific department or process where LLMs have driven significant improvements, detailing the challenges faced before implementation and the subsequent benefits realized. These narratives humanize the technology, illustrating its potential through relatable examples that resonate with stakeholders. Personal stories of employee impact and customer satisfaction further enrich the narrative, offering a glimpse into the real-world implications of LLM-driven innovation. By anchoring your presentation in storytelling, you create an emotional connection with your audience, enhancing the persuasiveness and memorability of your message.

Establishing a feedback loop with stakeholders is integral to refining your ROI communication strategies and ensuring their ongoing effectiveness. Regular stakeholder surveys can provide valuable insights into how well your presentations resonate with different audiences, highlighting areas for improvement and adaptation. These surveys serve as a barometer for gauging stakeholder engagement and understanding, allowing you to tailor your approach to meet their needs better. Iterative improvements based on stakeholder feedback ensure that your communication strategies remain dynamic and responsive, fostering a culture of continuous learning and enhancement. By actively seeking and incorporating stakeholder feedback, you demonstrate a commitment to transparency and collaboration, reinforcing trust and credibility in your ROI presentations.

As we conclude this chapter, the importance of a tailored communication strategy in presenting the ROI of LLM investments becomes abundantly clear. By customizing your approach to address the diverse needs of stakeholders, utilizing visual aids to enhance understanding, and harnessing the power of storytelling to create compelling narratives, you lay the foundation for persuasive and impactful presentations.

Establishing a feedback loop ensures that your communication strategies remain relevant and practical, fostering ongoing engagement and support from stakeholders. With these tools, you are well-equipped to articulate the value of LLMs within your organization, paving the way for successful integration and strategic growth.

Chapter Six:
Building the Team for LLM Success

*"Great things in business are never done by one person.
They're done by a team of people."*

– Steve Jobs, founder of Apple

Imagine a busy office where a team of professionals collaborates seamlessly, each member contributing their unique expertise to the successful deployment of Large Language Models (LLMs). The energy in the room is exciting as ideas flow freely, and innovations are born from the harmony of diverse talents. In this world, the roles required to implement LLMs are not merely positions on an organizational chart; they are essential to redefine how businesses operate and compete in the digital age.

As you consider the potential of LLMs for your organization, understanding the roles necessary to harness this technology is both a strategic imperative and an exciting opportunity to shape the future of your business.

1. Team Roles

At the helm of any LLM initiative is the AI Project Sponsor, a role that encapsulates the vision and strategic direction of the project. This individual is responsible for articulating the business problem that the AI is meant to solve, ensuring that the deployment aligns with organizational objectives and delivers tangible value. The AI Project Sponsor is not merely a figurehead but a pivotal leader who navigates the complexities of budget approvals, timeline management, and resource allocation. Their keen understanding of AI's impact on ROI, customer experience, and product value makes them an indispensable asset to the organization. They are the bridge between the executive suite and the operational teams, ensuring that AI initiatives are not only ambitious but also grounded in practical realities.

Supporting the AI Project Sponsor is the Product Manager, a role that serves as the linchpin between technical teams and business goals. The Product Manager defines the AI feature requirements that drive product integration, working closely with AI engineers, UX designers, and customer support teams to ensure a seamless deployment. This role demands a deep understanding of the technical and business aspects of LLMs and the ability to communicate effectively across diverse teams. By translating complex technical specifications into actionable business strategies, the Product Manager ensures that AI solutions meet the specific needs and objectives of the organization.

Central to the implementation of LLMs are the AI Engineers, the architects of the AI models that will transform business operations. These professionals are responsible for building, training, and fine-tuning AI models to ensure that they meet business goals and deliver optimal performance. AI Engineers work with software developers to deploy and maintain these models, ensuring they integrate seamlessly with existing systems and workflows. Their expertise in machine learning and data science is complemented by a deep understanding of natural language processing and LLM technologies, enabling them to create innovative and practical solutions. As the demand for AI expertise continues to grow, these engineers are among the most sought-after professionals in the tech industry, commanding high salaries and playing a critical role in the success of AI initiatives.

Data Scientists are the custodians of the data that fuels LLMs, tasked with collecting, cleaning, and preparing datasets for AI training. Their role is crucial for ensuring that AI models are trained on high-quality, unbiased data, as the integrity of the data directly impacts the performance and reliability of the AI. Data Scientists manage databases and AI pipelines, working closely with AI Engineers to ensure that the data is processed efficiently and effectively. Their analytical skills and attention to detail are essential in identifying patterns and insights to inform strategic decisions and drive business growth.

The role of the Software Developer or AI Integration Engineer is to embed AI models into existing software, applications, or websites, ensuring that they connect seamlessly with databases, APIs, and cloud services. This integration is vital for the functionality and scalability of AI solutions, as it allows organizations to leverage AI capabilities across various platforms and touchpoints. The AI Integration Engineer works closely with IT teams to ensure that AI solutions align with existing workflows, such as CRM systems, e-commerce platforms, and automation tools. This role ensures that AI initiatives deliver maximum value and efficiency by facilitating the smooth integration of AI into business operations.

The User Interface Designer plays a critical role in ensuring that AI-driven features are intuitive and user-friendly. Whether designing chatbots, recommendation engines, or automation tools, the User Interface Designer focuses on creating seamless and engaging interactions for customers. Their expertise in user-centric design principles ensures that AI solutions enhance the customer experience, driving satisfaction and loyalty.

The AI Ethics Officer has a role of growing importance, ensuring that AI initiatives adhere to ethical standards and legal requirements, such as the General Data Protection Regulation (GDPR) and the California Consumer Privacy Act (CCPA). This Officer monitors developing systems for AI bias and transparency, ensuring that AI solutions are fair, accountable, and aligned with organizational values. As the ethical implications of AI become increasingly scrutinized, the AI Ethics Officer plays a vital role in safeguarding the integrity and trustworthiness of AI deployments.

Finally, the Customer Support team is essential in training employees on the effective use of AI, handling customer feedback on AI-driven features, and adjusting implementations based on user needs. Their role is critical in bridging the gap between technology and end-users, ensuring that AI solutions are not only functional but also accessible and beneficial to all stakeholders.

For very small businesses where resources are constrained, a lean team can still achieve significant success in LLM initiatives. A Business Owner or Product Manager can provide strategic oversight, guide the project, and ensure alignment with business objectives and ethical standards. An AI/ML Engineer can focus on AI model development, while a Software Developer can handle AI integration. A Customer Support or Operations Lead can manage training and feedback, ensuring that AI solutions meet the needs of users and customers alike.

The skills and competencies required for these roles are multifaceted, encompassing proficiency in data analysis and machine learning, understanding of NLP and LLM technologies, and strong problem-solving and critical thinking abilities. Cross-functional collaboration is essential, with AI Liaisons facilitating communication between teams and Business Analysts with AI expertise providing valuable insights into project goals and outcomes.

Recruiting and onboarding talent for LLM initiatives require strategic approaches, leveraging AI talent networks and partnerships to attract skilled professionals. Onboarding programs tailored to LLM-specific roles can accelerate integration and productivity, ensuring that new hires are equipped to contribute effectively from the outset. By encouraging a collaborative and innovative environment, organizations can build the teams necessary to drive LLM success and achieve strategic objectives.

Interactive Element: Roles and Skills Checklist

To ensure your organization is well-prepared for LLM implementation, consider conducting a roles and skills assessment using the following checklist:

1. **Identify Key Roles:** Determine which roles are essential for your LLM initiative, considering the size and scope of your organization.

2. **Assess Skills and Competencies:** Evaluate the specific skills and competencies required for each role, focusing on data analysis, machine learning, and NLP expertise.

3. **Foster Cross-Functional Collaboration:** Ensure that roles are aligned with cross-functional collaboration requirements, facilitating communication and knowledge sharing.

4. **Develop Recruitment Strategies:** Leverage AI talent networks and partnerships to attract skilled professionals, and design onboarding programs tailored to LLM-specific roles.

5. **Assess inhouse talent:** Consider the staff you currently employ and determine which ones may fulfill a given role within the AI integration team. This will help identify which training programs you need.

By completing this checklist, you will gain a better understanding of your organization's readiness for LLM implementation and identify areas for enhancement.

2. Upskilling the Workforce: Training and Development

Within your organization, implementing an LLM requires a workforce that is not only familiar with the technologies but proficient in their application. This new initiative begins with designing comprehensive training programs that equip your current employees with the skills they need to thrive in an AI-driven environment.

Workshops on LLM fundamentals and applications serve as the foundation, providing employees with a thorough understanding of these models and their potential impact on business processes. These interactive workshops should encourage participants to engage with the material and apply their learning in practical scenarios. Such sessions can demystify complex concepts, transforming apprehension into curiosity and competence.

Hands-on training sessions are indispensable for solidifying this newfound knowledge. Allowing employees to interact directly with AI tools and platforms facilitates a deeper understanding of how these technologies function in real-world contexts. Employees should experiment with AI-driven applications, whether through building simple models or analyzing data sets, fostering an environment where learning by doing is the norm. This experiential approach reinforces theoretical concepts and builds confidence, enabling your staff to tackle AI challenges with assurance. Continuous learning modules focusing on emerging AI trends can enhance this process by keeping your workforce abreast of the latest advancements and innovations. These modules ensure that learning is not a one-time event but an ongoing journey, with employees encouraged to explore new ideas and technologies as they develop.

Certification and professional development are also key components of a successful upskilling strategy. By promoting certification programs in machine learning and data science, you provide your employees with a structured pathway to deepen their expertise in AI technologies. These credentials enhance their skill sets and testify to their proficiency, fostering a sense of pride and accomplishment. AI-focused leadership development programs can further empower your team by equipping them with the skills to lead AI initiatives effectively. These programs can help cultivate a new generation of leaders who understand the strategic implications of AI and can drive innovation within your organization.

Mentorship and peer-learning initiatives are crucial for fostering a culture of knowledge-sharing and skill development. By pairing experienced AI professionals with newcomers, you create an environment where expertise is transferred organically, allowing less experienced employees to learn from the insights and experiences of their mentors. This symbiotic relationship benefits both parties, as mentors refine their teaching and leadership skills while mentees gain valuable insights and guidance. Establishing learning circles for collaborative problem-solving further enhances this process, allowing employees to discuss challenges, share ideas, and develop solutions collectively. These circles can serve as incubators for innovation, as diverse perspectives come together to tackle complex issues and generate creative solutions.

Technology is a powerful ally in the pursuit of scalable training solutions, offering a myriad of tools and platforms that can facilitate learning across a distributed workforce. Online learning platforms, rich with AI and LLM content, allow employees to learn at their own pace, accessing resources and courses that align with their individual needs and interests. These platforms can be customized to reflect your organization's specific objectives and competencies, ensuring that learning is both relevant and impactful. Virtual reality (VR) simulations present another exciting avenue for real-world practice scenarios, immersing employees in interactive environments where they can apply their skills in a controlled, yet dynamic setting. These simulations can replicate complex business situations, allowing employees to experiment and learn without the fear of real-world consequences.

In implementing these training and development strategies, it's crucial to recognize that upskilling is not merely about imparting technical skills but also about fostering a mindset of adaptability and growth. Employees should be encouraged to embrace change, viewing it as an opportunity for personal and professional development rather than a challenge to be feared. By celebrating learning achievements and recognizing those who exhibit initiative and curiosity, you can reinforce a culture of continuous improvement where employees are motivated to excel and contribute to your organization's success.

One must not overlook the importance of setting clear goals and expectations for these training programs. Employees should understand what they are learning and why it is relevant to their roles and the broader organizational objectives. By aligning training with business goals, you create a sense of purpose and direction, ensuring that employees are not simply acquiring skills in a vacuum but are developing competencies that will drive tangible business outcomes. This alignment can also facilitate the measurement of training effectiveness, providing insights into how these programs contribute to your organization's strategic objectives.

The success of your upskilling initiatives hinges on a commitment to fostering a learning culture that values exploration, innovation, and collaboration. By investing in your employees' development, you enhance their skill sets and fortify your organization against the challenges and uncertainties of an AI-driven world. As your workforce becomes more adept at working through the complexities of LLM technologies, they will be better equipped to drive innovation, optimize processes, and deliver exceptional value to your customers.

3. Crafting a Learning Culture: Fostering Innovation

For organizations poised to integrate Large Language Models (LLMs), the ability to learn, adapt, and innovate becomes a cornerstone of sustained success. This ability begins with fostering an environment that tolerates and actively encourages experimentation and innovation. Leadership must champion a mindset where taking calculated risks is celebrated, and failures are seen not as setbacks but as learning opportunities. Recognizing and rewarding learning achievements creates a culture where intellectual curiosity and professional development are ingrained into the organizational culture, motivating employees to push boundaries and explore new possibilities.

Organizations must go beyond traditional methods to ignite innovation, allocating resources specifically for innovation labs and projects. These dedicated spaces serve as incubators for creative ideas, where teams are encouraged to test hypotheses, develop prototypes, and iterate on solutions in a low-risk environment. Hosting hackathons and innovation challenges further galvanizes creative thinking, bringing together diverse teams to tackle complex problems with fresh perspectives. Such events foster a spirit of camaraderie and collaboration while surfacing novel ideas that might otherwise remain dormant. By embedding these practices into the organizational ethos, you cultivate an environment where personal creativity flourishes, and groundbreaking solutions emerge organically.

Cross-departmental collaboration is another critical aspect of fostering innovation. By encouraging knowledge exchange across different departments, you break down silos and facilitate the flow of information that is vital for holistic problem-solving. Cross-functional teams working on AI projects bring together diverse skill sets and perspectives, enabling a more comprehensive approach to tackling challenges. Regular knowledge-sharing sessions and workshops provide opportunities for employees to share insights, learn from one another, and co-create solutions. This collaborative approach enhances the quality of outcomes and builds a sense of collective ownership and accountability, empowering employees to contribute meaningfully to the organization's strategic objectives.

Aligning learning objectives with business goals is paramount to ensuring that development efforts are both relevant and impactful. By defining clear learning goals linked to LLM outcomes, you provide employees with a roadmap that connects their personal development to the broader organizational mission. These goals should be specific, measurable, and aligned with the strategic priorities of the organization, enabling employees to track their progress and understand the value of their contributions. Integrating learning metrics into business performance reviews further reinforces this alignment, offering a tangible framework for assessing the impact of learning initiatives on business performance. This approach to performance reviews enhances accountability and ensures that learning remains a strategic enabler of organizational success.

In crafting a learning culture that fosters innovation, it's essential to recognize the role of leadership in modeling the behaviors and values you wish to see in your organization. Leaders must embody a growth mindset, demonstrating a commitment to continuous learning and embracing change as a catalyst for growth. By prioritizing their development and remaining open to new ideas, leaders set the tone for the rest of the organization, inspiring others to follow suit. This leadership style drives innovation and cultivates an environment where employees feel empowered to take ownership of their learning and contribute to the organization's success.

As you consider the next steps in fostering a culture of innovation, remember that the journey is ongoing and requires a sustained commitment to learning, collaboration, and creativity. By embedding these principles into the core of your organization, you position yourself not only to succeed in the face of ongoing change but to thrive, transforming challenges into opportunities and setting the stage for a future where AI and human ingenuity work hand in hand to achieve extraordinary results.

Chapter Seven:
Strategic Incremental Integration

"It is better to take many small steps in the right direction than to make a great leap forward only to stumble backward."

— Louis Sachar, American author

Imagine stepping into the executive suite of a multinational corporation where the air buzzes with anticipation and the promise of transformation. Senior engineers are poised to embark on a journey to integrate Large Language Models (LLMs) into their business processes. The stakes are high, and the potential rewards are even higher. This moment is where strategic foresight meets technological expertise, and the software engineering team, together with the AI team, translates the theoretical promise of AI into tangible business outcomes. In this chapter, I discuss the intricate process of AI integration, something that requires precision, coordination, and an unwavering focus on long-term goals.

1. Integration Framework

Much like any robust software development process, the integration framework for AI unfolds through multiple increments and multiple phases. These phases are not mere steps but dynamic cycles that evolve, adapt, and iterate over time. At the heart of this process lies the philosophy of breaking down long-term aspirations into manageable business objectives, each serving as an increment in an iterative cycle of development and refinement. This approach mirrors the principles of agile development, where flexibility and responsiveness are paramount, allowing for adjustments based on real-time feedback and changing market conditions.

By adopting this framework, you ensure that your AI initiatives remain aligned with overarching business goals, providing a roadmap that guides your organization through the complexities of AI integration with clarity and purpose.

In the initial stages of integration, it is crucial to establish a clear vision of what you aim to achieve with AI, articulating a strategic narrative that resonates across the organization. This vision serves as a North Star, guiding every decision and action, ensuring that all stakeholders are aligned and committed to the success of the initiative. As you embark on this journey, consider the unique challenges and opportunities that AI presents, recognizing that its transformative potential lies not only in its technological capabilities but also in its ability to reshape organizational culture, processes, and customer experiences. By fostering a culture of innovation, collaboration, and continuous learning, you create an environment where AI can thrive, driving meaningful change and delivering measurable value.

As you progress through the integration framework, it is essential to adopt a holistic approach, considering the interplay between data, algorithms, and infrastructure. These three pillars form the foundation of any successful AI initiative, each requiring careful planning and execution. A robust data strategy ensures access to high-quality, relevant data, the lifeblood of AI models, enabling them to learn, adapt, and improve over time. The algorithm strategy focuses on developing effective AI models, leveraging the expertise of data scientists and engineers to design robust and efficient systems. Meanwhile, the infrastructure strategy ensures that the computational power and hosting options are in place, providing the scalability and flexibility needed to support AI deployment at scale. By addressing each of these pillars with precision and foresight, you lay the groundwork for a seamless and effective AI integration, positioning your organization for long-term success.

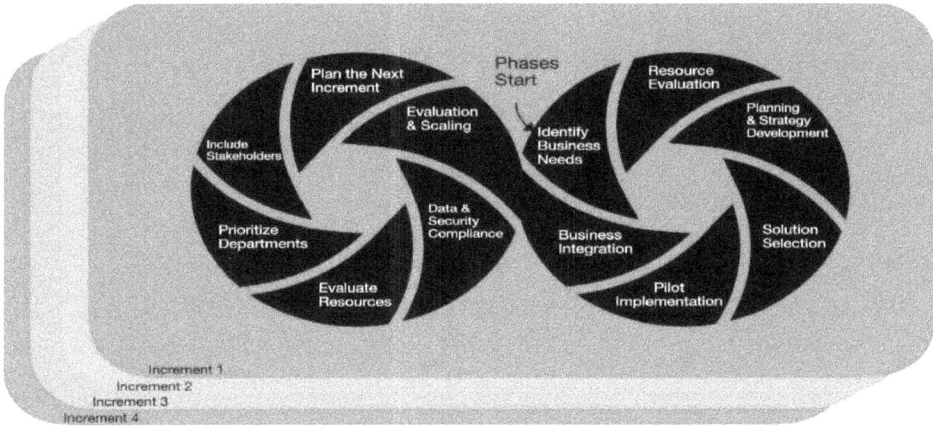

Incremental Integration

The iterative nature of the integration framework necessitates a commitment to continuous improvement, where each phase builds upon the insights and lessons learned from the previous cycle. This iterative process fosters a culture of experimentation and innovation, encouraging teams to explore new ideas, test hypotheses, and refine their approaches based on empirical evidence. By embracing this mindset, you cultivate an organization that is agile, adaptable, and resilient, capable of simplifying the complexities of AI challenges with confidence and clarity. As you advance through the integration process, remember that the journey is not a linear path but a dynamic, evolving cycle that requires constant reflection, adaptation, and recalibration.

Interactive Element: Integration Readiness Checklist

To assess your organization's readiness for AI integration, consider the following checklist:

1.   Have you established a clear vision and strategic objectives for AI integration?

2.   Is there a robust data strategy in place ensuring access to high-quality, relevant data?

3.   Are there effective algorithm strategies and expertise to develop powerful AI models?

4.   Does your infrastructure strategy support scalability and flexibility for AI deployment?

5.   Is there a culture of innovation, collaboration, and continuous learning within your organization?

6.   Have you adopted an iterative approach, allowing for continuous improvement and adaptation?

Reflecting on these questions will help you identify areas of strength and opportunities for growth, ensuring that your AI integration efforts are poised for success.

2. Increment 1: Phase 1: Identify Business Needs

For strategic AI implementation, the initial phase of business need identification is the cornerstone upon which all subsequent actions are built. This phase demands a meticulous audit of current pain points and inefficiencies, a process that begins with documenting every existing business process, no matter how minute, together with the data flow, in and out. Include all potential stakeholders in workshops so that no one is missed and all roles are defined. By creating a comprehensive map of workflows and identifying bottlenecks, you unveil areas where productivity falters, and inefficiencies loom large. Measuring the time consumed by repetitive tasks and pinpointing activities prone to error are crucial steps in this diagnostic journey, providing tangible evidence of where AI could serve as a transformative force. This analysis extends to calculating operational costs down to the last cent, enabling a clear understanding of financial outlays. Through this lens, specific processes ripe for AI enhancement come into sharp focus, revealing opportunities to streamline operations and drive efficiencies that align seamlessly with your organizational goals.

To aid this investigative process, here are some insights that are true today. Humans are better than computers for:

- Creative problem-solving, such as brainstorming or generating new ideas
- Emotional intelligence, such as empathy
- Social interaction, such as building relationships or negotiation
- Making ethical decisions, such as those decisions requiring a moral component
- Handling ambiguity, such as making judgments in complex situations that require context
- Physical dexterity, where nimble movements are required

Alternatively, computers are better than humans for:

- Large-scale data analysis, such as pattern recognition
- Repetitive tasks, such as making routine actions with accuracy and speed
- Complex mathematical computations, such as solving equations
- Information retrieval, such as finding and displaying data
- Quality control checks, such as finding errors in standard processes

Apply these insights to your task analysis, and the processes appropriate for automation will be apparent.

With a clear understanding of the current business marketplace, the next step involves conducting a thorough impact analysis. This is where the potential of AI begins to take shape in quantifiable terms as you project the potential impact of proposed AI interventions on your business operations. The goal is to quantify the potential return on investment (ROI), transforming abstract possibilities into concrete figures that can galvanize stakeholder support. This task involves estimating direct financial gains and considering indirect benefits such as enhanced employee satisfaction and strengthened brand reputation. By projecting these outcomes, you create a compelling narrative that underscores the strategic value of AI, paving the way for informed decision-making. This comprehensive analysis forms the bedrock of a strategic AI initiative, ensuring that every step taken is rooted in a deep understanding of both the challenges and opportunities that lie ahead.

3. Phase 2: Resource Evaluation

As we continue to explore the strategic implementation of AI, the second phase, resource evaluation, demands a meticulous examination of your organization's technical infrastructure. This infrastructure is the backbone that supports the deployment of Large Language Models, and a solid foundation is critical for success. The first step in this assessment involves taking inventory of your current technological resources, identifying any gaps that may hinder the effective integration of AI systems. This task requires a deep dive into your existing hardware capabilities, ensuring that your infrastructure can accommodate the computational demands of LLMs. The processing power needed for these models often necessitates upgrading to more advanced servers or adopting cloud-based solutions, which offer scalability and flexibility.

Moreover, evaluating your network architecture is crucial, as the seamless operation of AI systems depends on robust connectivity and data flow. Consider the bandwidth and latency requirements to ensure that your network can handle the increased data traffic that accompanies AI integration. This analysis should extend to data storage solutions, as LLMs require substantial data for training and operation. With their inherent scalability, cloud storage options often present an attractive choice, allowing for real-time data access and collaboration across dispersed teams. Additionally, you must address cybersecurity measures, safeguarding sensitive data from breaches and ensuring compliance with regulatory standards.

Finally, the assessment must encompass software tools and platforms that facilitate AI model development and deployment. Investigate whether your current software environment supports the necessary AI frameworks and libraries, such as TensorFlow or PyTorch, which are instrumental in training and deploying LLMs. This phase is not merely about identifying deficiencies but also about planning for future needs, anticipating advancements in AI technology, and ensuring that your infrastructure can evolve to meet these demands.

4. Phase 3: Planning and Strategy Development

The complexity of AI integration requires a meticulously crafted plan where objective setting lays the groundwork for success. Defining specific, measurable goals ensures that the AI initiative is not only aspirational but also actionable. These goals should align with broader business objectives, capturing immediate improvements and long-term aspirations. Realistic timelines are crucial, balancing ambition with practicality and providing a roadmap for the phased implementation of AI systems. Performance benchmarks serve as signposts along this journey, allowing for the assessment of progress and the calibration of strategies as necessary. Establishing budget parameters is equally vital, as well as ensuring that financial resources are allocated judiciously to support the AI initiative without straining the organization's overall fiscal health.

Risk management is the bedrock of a robust AI strategy, addressing potential challenges before they escalate into crises. Data security concerns loom large, necessitating stringent measures to protect sensitive information from breaches and unauthorized access. Compliance with privacy regulations, such as the General Data Protection Regulation (GDPR), a European Union law adopted in 2016, or the California Consumer Protection Act (CCPA) from 2018, is non-negotiable, requiring a comprehensive understanding of legal obligations and the implementation of policies that safeguard user data. System downtime, a perennial concern, must be mitigated through redundancy and failover strategies, ensuring continuous operation despite technical disruptions.

Employee resistance, often stemming from fear of job displacement or change, requires proactive engagement and transparent communication to foster buy-in and alleviate apprehensions. Integration challenges, from technical incompatibilities to workflow disruptions, demand a flexible and adaptive approach, one that leverages cross-functional collaboration and iterative refinement to ensure a seamless transition. These risks can be effectively managed through careful planning and strategic foresight, setting the stage for a successful AI implementation that aligns with the organization's vision and values.

5. Phase 4: Solution Selection

As you navigate the intricacies of AI integration, the decision to either buy or build your AI solution takes center stage. This choice requires a nuanced understanding of your specific needs and strategic vision. Designing a high-level module that aligns perfectly with your business objectives is often the starting point. This task involves engineering an API that seamlessly integrates with your existing systems, ensuring compatibility and functionality. Estimating the cost to build this proprietary solution is crucial, as it provides a baseline against which vendor offerings can be evaluated. Setting this benchmark allows you to compare the potential benefits and drawbacks of developing an in-house solution versus purchasing from an external vendor. The decision-making process is multifaceted, requiring a thorough analysis of several factors.

Vendor evaluation is a critical component of this phase. It demands a comprehensive cost comparison, weighing the financial implications of each option. Beyond mere pricing, detailed feature analysis is essential to ensure the vendor solution meets your operational needs. Support services and implementation assistance that vendors provide play a pivotal role in seamlessly integrating AI into your business processes. Additionally, assessing a vendor's track record and reliability can offer insights into their ability to deliver on promises. A vendor's history of successful implementations and customer satisfaction indicates their capability to support your AI initiatives.

Integration requirements further complicate the decision. API compatibility is paramount, ensuring the new solution effectively communicates with your existing systems. Data migration needs should be assessed to facilitate a smooth transition, minimizing disruptions. System updates and user access management must also be considered to maintain operational integrity and security. Ultimately, the decision to buy or build hinges on carefully examining cost differences, time constraints, and the long-term implications on competitive advantage and scalability. Complete control over a custom-built solution offers flexibility and alignment but may come with higher initial costs and development time. Conversely, a vendor solution may provide faster deployment and proven reliability but could limit customization. Additionally, evaluating vendor stability and the potential for market changes ensures that your chosen solution remains viable. Last but not least, do not forget maintenance needs as they impact the ongoing sustainability of the solution.

6. Phase 5: Pilot Implementation

As you embark on the pilot implementation phase, it's crucial to identify a project that balances low risk and high impact, ensuring that the initiative is manageable and significant in its potential benefits. This selection process involves setting clear boundaries that delineate the scope of the pilot, preventing scope creep, and maintaining focus on the core objectives. Success metrics must be defined with precision, providing a framework against which the pilot's outcomes can be measured. These metrics, whether they pertain to efficiency gains, cost reductions, or customer satisfaction improvements, serve as the benchmarks for evaluating the pilot's effectiveness. Alongside these metrics, the creation of rigorous testing protocols is essential for the testing manager to ensure thorough unit testing and end-to-end system evaluations.

Resource allocation is another critical facet of the pilot phase. Assigning a dedicated team is paramount, as it guarantees that the project receives the focused attention and expertise required for its success. This team should be equipped with budgetary allocations, providing the financial resources needed to execute the pilot with minimal constraints. Additionally, the project manager should establish a precise time commitment, setting realistic timelines that accommodate the complexities of AI integration. Supporting this team with a robust support system is equally essential as offering technical assistance and troubleshooting capabilities to address any challenges during the pilot.

Developing the software or vendor product is a dynamic process involving continuous iteration. Initial development efforts should focus on creating a working module that aligns with the project's objectives. This module must be subjected to rigorous testing at the unit level and within the broader context of the business process it aims to enhance. Iteration is key here, with each round of testing serving as an opportunity to refine and optimize the module, ensuring it meets the defined success metrics.

Once testing is complete, the operations supervisor can deploy the module, marking pa significant milestone in the pilot phase. However, the journey doesn't end with deployment; measuring success through the lens of the established metrics provides valuable insights into the module's performance and impact. An incremental adoption strategy is advisable, allowing for a gradual rollout that minimizes disruption to existing operations, ensuring a smooth transition, and maximizing the pilot's potential for success.

7. Phase 6: Business Integration

Bringing Large Language Models into your business operations is akin to weaving a complex tapestry, where the project manager meticulously places each thread to achieve the desired pattern of success. At the core of this integration lies a robust communication strategy designed to maintain alignment and transparency across all stakeholders. Regular updates serve as the lifeblood of this strategy, ensuring that everyone, from executives to frontline employees, is aware of the progress and the transformative potential of AI initiatives. These updates should not merely recount milestones but clearly explain the benefits, reinforcing the narrative of how AI is poised to enhance efficiency, drive innovation, and create value. Sharing transparent timelines further solidifies trust, as it sets realistic expectations and offers a glimpse into the strategic roadmap guiding the integration. Open feedback channels, an often-overlooked element, are indispensable; they empower stakeholders to voice concerns, share insights, and contribute to the ongoing refinement of AI systems, fostering a culture of collaboration and continuous improvement.

Parallel to the communication efforts, a comprehensive training program is essential to equip your workforce with the skills needed to leverage AI effectively. The journey begins with a skill assessment, identifying gaps and strengths within your team, which informs the development of tailored training materials. These resources should cater to various learning preferences, incorporating visual aids, hands-on exercises, and theoretical content to ensure a well-rounded understanding of AI technologies. Scheduling workshops at convenient intervals allows employees to engage with the material without disrupting their daily responsibilities, promoting a seamless learning experience. Progress-tracking mechanisms are vital, offering insights into individual and collective advancements and enabling targeted interventions where necessary. Support documentation complements this training, providing a reference point for employees learning the nuances of integration. This holistic approach to training not only empowers employees but also instills confidence, ensuring that they are not just passive participants but active contributors to the AI-driven transformation of your organization.

8. Phase 7: Evaluation and Scaling

As we transition into your AI integration's evaluation and scaling phase, the focus shifts to rigorous metric tracking, a critical process ensuring your AI systems' effectiveness and viability. Performance improvements, often the first indicator of success, must be meticulously measured, capturing speed, accuracy, and efficiency enhancements. This measurement must extend to cost savings, where tangible reductions in operational expenses can significantly impact your bottom line. User adoption rates, a testament to the system's usability and acceptance, provide insights into how well the AI solution has been integrated into daily operations. Error reduction further underscores the AI's capability to minimize inaccuracies and enhance reliability, while time savings highlight the operational efficiencies gained through automation. Lastly, the impact on customer satisfaction cannot be overlooked, as AI-driven improvements in service and product offerings are pivotal to enhancing the overall customer experience.

Feedback collection is an equally vital component of this phase, serving as a conduit for continuous improvement and refinement. User surveys, a direct line to those interacting with the AI systems, offer qualitative insights that can guide future developments. System performance data, rich with quantitative metrics, objectively assesses the AI's functionality and areas for enhancement. Process efficiency metrics further illuminate the effectiveness of workflow optimizations, offering a clear picture of the AI's impact on operational processes. A comprehensive cost-benefit analysis, capturing both the tangible and intangible benefits realized, completes the evaluation, providing a holistic view of AI's contribution to organizational success. This feedback loop is instrumental in informing strategic decisions, ensuring the AI systems remain aligned with evolving business objectives and continue delivering value as they scale across the organization.

9. Plan the Next Increment

Stakeholder inclusion at this stage is critical for success. They will be heavily invested in setting priorities. Revisiting the initial phase of business need identification, you find yourself at a crucial juncture where the groundwork laid by the pilot has set the stage for strategic scaling. This moment requires a meticulous approach to scaling strategy, where department prioritization becomes paramount. It's not just about identifying which departments stand to benefit most from AI integration but also about evaluating their readiness and strategic importance. Prioritizing departments means assessing where AI can drive the most significant value, considering factors such as the potential for cost savings, efficiency improvements, and alignment with long-term business objectives. This prioritization process ensures that resources are allocated effectively, maximizing the impact of AI deployment across the organization.

Resource requirements emerge as a focal point in this phase, demanding a comprehensive evaluation of what is necessary to support the next increment of AI integration. This phase involves a detailed analysis of human and technological resources, ensuring the infrastructure can support expanded AI capabilities while addressing any gaps hindering progress. Timeline development is equally critical, requiring a nuanced balance between speed and thoroughness. Setting realistic timelines involves considering the complexities of coordination across departments, the availability of key personnel, and the potential for unforeseen challenges. This careful planning ensures that the AI integration progresses smoothly without overwhelming existing operations or compromising quality.

Budget allocation is another vital element, necessitating a strategic approach to financial planning. This planning involves determining the overall budget for the next increment and ensuring that funds are distributed in a manner that aligns with strategic priorities. Risk assessment updates refine this process further, providing insights into potential pitfalls and helping to allocate resources to areas of greatest need. This iterative return to phase one is not merely a repetition but a refinement, drawing on the insights gained from previous phases to inform and enhance the next steps in your AI integration journey.

10.   Ongoing Planning

This phase is about maintaining current systems and ensuring they remain robust, secure, and efficient. Regular system updates are imperative, as they keep your AI infrastructure in tune with the latest advancements, preventing obsolescence and enhancing functionality. Performance monitoring is equally crucial, continuously checking the system's health and operational efficiency. It allows you to identify bottlenecks and optimize processes in real time, ensuring that your AI systems deliver consistent value.

Your Compliance Officer must rigorously enforce security protocols to protect sensitive data from breaches, a concern that becomes even more pressing in an era where cyber threats are increasingly sophisticated. This task involves implementing multi-layered security measures, including encryption, access controls, and regular security audits. Alongside this, establishing comprehensive backup procedures is essential for safeguarding data integrity. These procedures ensure that critical information is preserved and recoverable in case of system failures, minimizing downtime and disruption. Emergency response plans further fortify your defenses, enabling swift action in the face of unforeseen challenges, thus ensuring business continuity and resilience.

Continuous improvement is the lifeblood of sustainable success. Regular performance reviews are not mere formalities but opportunities to assess progress, identify areas for enhancement, and recalibrate strategies. The implementation of updates, informed by these reviews, ensures that your systems evolve in alignment with business needs. New feature evaluation is ongoing, allowing you to leverage emerging technologies and capabilities to maintain a competitive edge. User feedback integration provides invaluable insights, transforming end-user experiences into actionable data that guides iterative development.

Process optimization, the final pillar, focuses on streamlining operations, eliminating inefficiencies, and enhancing productivity. This relentless pursuit of excellence, driven by a commitment to continuous improvement, ensures that your AI systems remain agile, relevant, and impactful in an ever-changing business marketplace.

11. Case Studies: Successful Strategic Implementations

Across diverse sectors, the strategic implementation of Large Language Models (LLMs) has catalyzed remarkable transformations, offering both challenges and opportunities unique to each industry. For instance, companies in retail have leveraged LLMs to refine inventory management and enhance demand forecasting. By analyzing purchasing trends and consumer behavior, LLMs have enabled retailers to optimize stock levels, reducing overstock and minimizing shortages, thereby aligning inventory with real-time market demand. This strategic approach improved operational efficiency and maximized profitability by tailoring inventory to consumer needs.

In education, integrating LLMs has revolutionized personalized learning experiences and curriculum adaptation. Educators have utilized these models to assess individual learning styles and preferences, crafting tailored educational pathways that enhance student engagement and achievement. The ability to adapt curricula dynamically based on real-time feedback has empowered educators to address diverse learning needs, fostering an inclusive and adaptive educational environment that promotes student success.

A foray into the real estate sector reveals LLMs' potential in streamlining processes and enhancing client interactions. Real estate professionals have transformed how they engage with clients by automating property descriptions and analyzing market trends, providing timely, relevant information that facilitates decision-making and enhances customer satisfaction. The strategic application of LLMs in real estate has optimized operational workflows and elevated the overall client experience.

Healthcare, with its complexity and demand for precision, has seen LLMs facilitate predictive analytics and patient management. By analyzing patient data, LLMs have assisted in diagnosing conditions and personalizing treatment plans, improving patient outcomes, and streamlining administrative processes. This strategic deployment has reduced bottlenecks in patient care, leading to more efficient healthcare delivery and heightened patient satisfaction.

These case studies underscore the versatility and transformative potential of LLMs across industries. As we transition to the next chapter, we'll explore the ethical considerations and governance structures necessary to ensure responsible AI deployment.

Chapter Eight:
Privacy, Data Quality, and Security

"Privacy is not an option, and it shouldn't be the price we accept for just getting on the Internet."

– Gary Kovacs, CEO of Accela

Imagine, if you will, a renowned global enterprise on the cusp of unveiling its latest AI-driven customer engagement platform. As the launch date draws nearer, the executive team convenes to tackle a crucial aspect underpinning the entire initiative: data privacy. The stakes are extraordinarily high for companies embracing AI technologies, such as Large Language Models (LLMs). The ability to protect personal data not only ensures compliance with stringent legal frameworks but also fortifies consumer trust, offering a competitive edge that can distinguish an organization in a crowded marketplace. Indeed, a robust data privacy strategy can transform potential vulnerabilities into pillars of strength, supporting ethical governance and sustainable growth. The importance of data privacy in AI applications cannot be overstated.

1. Privacy and Protection

In an era where data breaches can erode consumer confidence and devastate reputation capital, the necessity for strong data protection measures becomes unequivocal. The concept of consumer trust has evolved into a formidable competitive advantage that can propel a brand to new heights or, conversely, plunge it into obscurity. Regulatory standards, such as the General Data Protection Regulation (GDPR) in Europe and the California Consumer Privacy Act (CCPA), impose rigorous standards for data handling, demanding transparency, accountability, and security in AI deployments. These regulations establish legal parameters and set a benchmark for ethical AI practices, compelling organizations to integrate privacy considerations from the outset of system design. This practice ensures that data protection is not an afterthought but a foundational element of AI strategy.

Various data anonymization techniques are among the tools at your disposal for safeguarding data privacy. These methods obscure personal identifiers, allowing any analysis to use the data without compromising individual privacy. Pseudonymization is one such technique where private identifiers are replaced with pseudonyms, maintaining the integrity of the data while concealing its source. This approach is particularly valuable in data processing, as it enables the extraction of insights without direct exposure to identifiable information. Differential privacy, another advanced technique, adds controlled noise to datasets, ensuring that individual data points cannot be isolated while preserving overall data utility. This method provides a mathematical guarantee of privacy, making it a powerful tool for any organization committed to ethical data use. Data masking modifies data values, preventing sensitive information identification or reverse engineering. By implementing these techniques, organizations can strike a delicate balance between data utility and privacy protection, fostering an environment where technologists can collect insights without exposing personal data to undue risk.

In AI, privacy-preserving machine learning methods have emerged as vital innovations, enabling models to learn from data while upholding stringent privacy standards. Federated learning epitomizes this approach by decentralizing data processing, allowing AI models to train across multiple devices or servers without sharing raw data. This technique not only enhances privacy by keeping data local but also reduces the risk of data breaches by minimizing centralized data storage. Homomorphic encryption offers another layer of security, enabling computation on encrypted data without decrypting it, thus preserving privacy throughout the data processing pipeline. These methodologies are emblematic of a paradigm shift toward privacy-conscious AI, reflecting an industry-wide commitment to protecting individual rights while advancing technological capabilities.

Compliance with regional and international data protection regulations is not merely a legal obligation but a strategic necessity for businesses operating in the AI domain. The GDPR, for instance, profoundly impacts AI systems by imposing strict guidelines on data processing, transparency, and individual rights. Organizations must ensure that their AI applications adhere to principles such as data minimization, purpose limitation, and lawful processing while providing individuals with access to their data and the ability to contest automated decisions. The GDPR mandates privacy-preserving technologies, requiring organizations to conduct Data Protection Impact Assessments (DPIAs) for high-risk AI projects and to demonstrate compliance through documented safeguards and accountability measures. Similarly, the CCPA underscores the importance of data privacy, granting consumers rights over their information and imposing penalties for non-compliance. Understanding and implementing these regulatory frameworks demands a proactive approach, where privacy considerations are embedded into the AI system design and deployment.

Interactive Element: Privacy Protection Checklist

To assess and enhance your organization's data privacy strategy, consider the following checklist:

1. Have you implemented pseudonymization, differential privacy, or data masking techniques to protect sensitive data?

2. Is federated learning or homomorphic encryption employed to ensure privacy-preserving machine learning?

3. Do you comply with GDPR and CCPA regulations, including conducting DPIAs for high-risk AI projects?

4. Do you have documented processes for obtaining and managing consumer consent for data processing?

5. Are privacy considerations integrated into your AI systems' design and development stages?

6. Have you established a robust framework for monitoring and auditing data privacy practices?

Reflecting on these questions will provide valuable insights into your organization's privacy protection measures and highlight areas for improvement.

As you learn more about the intricacies of data privacy, remember that the goal is not merely compliance but creating a culture of trust and transparency. By embedding privacy into your AI strategy, you protect your organization from potential legal repercussions and build a foundation of ethical integrity that resonates with consumers, partners, and stakeholders. With ever-evolving digital platforms, where data is both an asset and a liability, safeguarding privacy becomes a defining characteristic of successful, forward-thinking enterprises.

2. Ensuring Data Quality: The Foundation of Reliable AI

Large Language Models (LLMs) must use data quality mechanisms to provide the reliability and efficacy that users expect. Data quality encompasses several critical components, each contributing to the overall robustness of AI outputs. At the forefront is accuracy, which ensures that the data collected truly reflects the reality it aims to represent. Consistency follows closely, demanding that data be uniformly collected and processed, minimizing discrepancies that could lead to errant AI predictions. Equally important is the completeness of datasets, as incomplete data can result in significant gaps in

AI models' understanding and performance. Relevance is the last but not least element, ensuring that the data used is pertinent to the task at hand, thereby enhancing the applicability and precision of AI insights. Together, these components form the foundation for reliable AI systems, underscoring the critical importance of meticulous data management in AI projects.

Your Data Specialist must employ several strategies to maintain high data quality, each tailored to address the unique challenges and requirements of LLM projects. Data validation and cleansing processes are indispensable in identifying and rectifying datasets' inaccuracies. These processes often involve sophisticated algorithms capable of detecting anomalies and inconsistencies, ensuring that the data fed into AI models is accurate and consistent. Regular data audits and quality checks further bolster this effort, providing a systematic approach to monitoring data integrity over time. These audits identify current issues and preempt potential future problems, allowing for timely interventions that preserve data quality. Advanced data preprocessing techniques also play a vital role, employing methodologies such as normalization and transformation to prepare data for efficient and effective AI processing. These techniques enhance the quality and usability of data, ensuring that it can be seamlessly integrated into AI models, optimizing their performance and outcomes.

Understanding the provenance of data is another critical aspect of ensuring data quality, particularly in complex AI projects where datasets may be derived from multiple sources. Data provenance refers to documenting data origins and transformations, providing a transparent and traceable record of data's journey from collection to analysis. This traceability is paramount for several reasons. Firstly, it ensures data integrity, allowing organizations to verify the authenticity and reliability of their data. Secondly, it facilitates accountability, enabling stakeholders to track changes and modifications throughout the data lifecycle.

96

Documenting data sources and transformations is a key component of this process, as it provides a clear and comprehensive overview of how data has been collected, processed, and utilized. Tracking data lineage in complex datasets further enhances this transparency, offering insights into the interconnections and dependencies between different data elements. By maintaining detailed records of data provenance, organizations can ensure that their AI models are built on a solid foundation of trustworthy and verifiable data.

The impact of data quality on AI outcomes is profound, influencing the accuracy and reliability of LLM outputs and the overall success of AI initiatives. High-quality data serves as the lifeblood of AI systems, enabling them to deliver insights and predictions that are both accurate and actionable. Conversely, poor data quality can have detrimental effects, leading to erroneous conclusions and misguided decisions. Consider, for instance, a case study in the healthcare sector where an AI model was tasked with predicting patient outcomes based on historical medical data. The model initially produced suboptimal results, with predictions that failed to align with actual patient trajectories. Upon investigation, they discovered that the dataset was rife with inaccuracies and inconsistencies, undermining the model's ability to make reliable predictions. Once they addressed these data quality issues through rigorous validation and cleansing processes, the model's performance improved dramatically, producing predictions that closely matched real-world outcomes. This case illustrates the critical role of data quality in enabling AI systems to fulfill their potential, highlighting the need for organizations to prioritize data management as a core component of their AI strategies.

In another example, a retail company sought to enhance its customer recommendation engine using an LLM trained on a vast transaction and browsing data set. Initially, the recommendations generated were often irrelevant or inaccurate, leading to customer dissatisfaction. Upon conducting a thorough data quality assessment, it became evident that the dataset was incomplete, with significant gaps in transaction histories and customer preferences. By addressing these issues and ensuring the completeness and relevance of the data, the company was able to dramatically improve the performance of its recommendation engine, resulting in more personalized and accurate suggestions that enhanced customer satisfaction and engagement. This example underscores the importance of data quality in driving successful AI outcomes, demonstrating how high-quality data can transform AI systems from underperforming tools into strategic assets that deliver tangible business value.

As you navigate the complexities of AI implementation, you must recognize that data quality is not a static attribute but a dynamic process that requires ongoing attention and refinement. Regular audits and cleansing efforts, coupled with a commitment to understanding data provenance, are essential for maintaining the integrity and utility of AI systems. By embracing these practices, you can ensure that your organization is well-equipped to leverage the full potential of LLMs, driving innovation and achieving strategic objectives with confidence and precision.

3. Security Protocols for LLM Deployments

Security is critical when deploying Large Language Models (LLMs), necessitating a sophisticated understanding of the many threats accompanying these advanced AI systems. As a business executive, you are acutely aware that, while transformative, LLMs also introduce unique vulnerabilities that you must diligently address to protect your organization's digital assets. Among the most pressing threats are data breaches and unauthorized access, where malicious actors exploit weaknesses in security defenses to access sensitive information. These breaches compromise confidential data and erode trust, potentially leading to significant financial and reputational damage.

Adversarial attacks pose another formidable challenge, where attackers manipulate input data to deceive AI models, leading to erroneous outputs and potentially catastrophic consequences for decision-making processes. Such attacks exploit the very nature of LLMs, taking advantage of their complex algorithms and vast datasets to introduce subtle yet impactful perturbations that compromise model integrity.

A robust security framework is imperative to safeguard against these threats, encompassing a range of measures designed to fortify the infrastructure supporting LLM deployments. Network security protocols and firewalls serve as the first line of defense, establishing barriers restricting unauthorized access and monitoring traffic for suspicious activities. The security engineer must meticulously define, configure, and regularly update the security protocols to adapt to evolving threats, ensuring that your network remains impervious to intrusion attempts.

Secure API management practices further enhance this defense, providing controlled access to LLM functionalities and protecting against unauthorized exploitation of AI capabilities. As conduits for data exchange, APIs must be rigorously secured, employing authentication and encryption measures to prevent interception and tampering. Regular security audits and vulnerability assessments are essential components of this framework, offering a comprehensive evaluation of your security posture and identifying potential weaknesses before anyone can exploit them. The security analyst should conduct these assessments routinely, leveraging both automated tools and expert analysis to ensure a thorough examination of your LLM infrastructure.

Encryption is pivotal in securing LLM data at rest and in transit. End-to-end encryption methods provide a robust solution, ensuring that data is encrypted from the point of origin to its final destination, thereby protecting against interception and unauthorized access. This approach safeguards data during transmission and ensures that stored data remains secure, mitigating the risk of breaches. Secure key management practices are integral to this process, involving the generation, distribution, and storage of cryptographic keys that are used to encrypt and decrypt data.

These practices must adhere to stringent security standards, ensuring that keys are stored in secure environments and protected against unauthorized access or loss. By implementing these encryption techniques, you can assure stakeholders that data security is a top priority, fostering confidence and trust in your organization's AI initiatives.

Incident response and recovery strategies form a critical component of your security framework, providing a structured approach to managing and mitigating the impact of security incidents. Developing comprehensive incident response plans is essential, detailing the procedures and responsibilities for identifying, containing, and resolving security breaches. These plans should be tailored to your organization's needs, considering the unique risks and vulnerabilities associated with LLM deployments.

Regular security drills and simulations strengthen your incident response capabilities, allowing your team to practice and refine their response strategies in a controlled environment. These exercises enhance preparedness and identify potential gaps or weaknesses in your response plans, enabling continuous improvement.

Continuous monitoring for threat detection is another crucial element, employing advanced tools and technologies to provide real-time insights into your security environment. This proactive approach allows for the early identification of potential threats, enabling swift action to prevent or mitigate incidents before they escalate.

As for AI security, staying ahead of potential threats and vulnerabilities is a constant challenge, requiring a proactive and comprehensive approach to risk management. By understanding the unique security risks associated with LLM deployments and implementing robust security protocols, you can protect your organization's digital assets and ensure the integrity of your AI initiatives.

As we transition to the next chapter, we will explore the broader implications of these security measures, examining how they contribute to the overall success and sustainability of AI-driven business strategies.

Chapter Nine:
Integration Challenges

"Most software today is very much like an Egyptian pyramid with millions of bricks piled on top of each other, with no structural integrity, but just done by brute force and thousands of slaves."

– Alan Kay, computer scientist who led the development of Smalltalk at Xerox PARC

Imagine yourself at the helm of a thriving enterprise, poised to embark on a transformative journey with Large Language Models (LLMs) at the core of your strategic vision. The allure of AI beckons with promises of enhanced efficiency, innovation, and competitive advantage. Yet, as with any ambitious endeavor, the path to successful AI integration is fraught with potential pitfalls that can derail your efforts if not meticulously navigated. The journey from AI ambition to achievement is often littered with obstacles that may seem insurmountable but are, in reality, opportunities for growth and refinement.

1. Preparing for Change

One of the most pervasive challenges for businesses is the sheer complexity of integrating AI into existing systems. Many organizations find themselves grappling with outdated infrastructure that is ill-equipped to support the computational demands of sophisticated AI models. This technological inertia can severely impede the deployment and scalability of LLMs, rendering their potential benefits moot. A comprehensive audit of your current infrastructure is imperative to surmount this challenge. This evaluation should encompass not only the hardware and software capabilities but also the compatibility of existing systems with the requirements of AI applications. Investing in up-to-date, flexible infrastructure that can accommodate the evolving demands of AI will lay a solid foundation for successful integration.

Another significant hurdle is the scarcity of in-house AI and data science expertise, which can stymie progress and lead to suboptimal implementation. Many organizations rely on external consultants to fill this gap, yet this approach can inadvertently result in a dependency that undermines internal capacity building. The key to overcoming this challenge is fostering a culture of continuous learning and development within your organization. By investing in the upskilling of your workforce, you can cultivate a cadre of AI-literate employees who possess the requisite skills to manage and optimize AI systems. This path empowers your team to take ownership of AI initiatives and ensures that your organization is well-positioned to adapt to future technological advancements.

Data privacy and security concerns represent formidable obstacles in the path of AI integration. The deployment of LLMs necessitates extensive datasets, raising questions about protecting and ethically using sensitive information. Navigating the maze of data privacy regulations such as GDPR and CCPA requires a vigilant approach to data governance. Implementing robust data protection measures, including encryption and access controls, is essential to safeguarding the integrity and confidentiality of your data. Moreover, fostering a culture of transparency and accountability within your organization will build trust with stakeholders and ensure compliance with regulatory frameworks.

Determining intellectual property ownership of AI-generated content poses a complex challenge, particularly in industries where proprietary information is a key competitive differentiator. The ambiguity surrounding the legal status of AI-generated outputs can lead to disputes and potential infringements. It is crucial to establish clear policies and ownership agreements that delineate the rights and responsibilities of all parties involved in AI projects in order to mitigate the risk of disputes. Engaging legal experts with experience in AI and intellectual property law can provide valuable guidance in crafting contracts that protect your organization's interests while fostering innovation.

Technological overwhelm is another common pitfall, as the proliferation of AI tools and platforms can lead to decision paralysis. With many options available, selecting the most appropriate tools for your organization's specific needs can be daunting. The temptation to adopt the latest and most advanced technologies can be compelling, yet it is essential to approach AI tool selection with a strategic mindset. Begin by identifying the specific business challenges you aim to address with AI and evaluate tools based on their ability to meet these objectives. Prioritizing solutions that integrate seamlessly with your existing workflows and deliver measurable value will ensure a more focused and effective implementation.

Gaining customer acceptance of AI-driven solutions can also be a formidable challenge, particularly in industries where human interaction is paramount. Data privacy and job security concerns can engender skepticism and resistance among customers and employees. It is critical to engage in open and transparent communication about the benefits and limitations of AI to overcome resistance barriers. Educating stakeholders on the ethical considerations and safeguards in place to protect their interests can alleviate fears and foster a more receptive environment for AI adoption.

The challenge of balancing AI's capabilities with human input is particularly pronounced when leveraging LLMs for content creation. While these models excel at generating text, the absence of a human touch can result in outputs that lack personalization and nuance. It is crucial to foster collaboration between AI and human contributors, ensuring that AI-generated content is refined and contextualized by human expertise. This symbiotic relationship between AI and human input can enhance the quality and relevance of content, delivering a more engaging and authentic experience for your audience.

The tendency to rush AI implementation without a clear strategic framework can lead to fragmented efforts and disjointed outcomes. Implementing AI without a comprehensive understanding of its implications can result in initiatives that fail to align with broader business objectives. Developing a robust AI strategy that articulates clear goals, success metrics, and implementation timelines is essential. Assessing your organization's strengths, weaknesses, and opportunities should thoroughly inform this strategic blueprint, ensuring that AI initiatives align with your overarching vision and objectives.

Resistance to change is an inherent human trait that can impede the adoption of new technologies. Employees accustomed to established processes may be hesitant to embrace AI-driven solutions, perceiving them as threats to their roles and responsibilities. Overcoming this resistance requires a concerted effort to engage and empower your workforce. You can foster a culture of innovation and collaboration by involving employees in the AI adoption process and providing them with the necessary training and support. Encouraging employees to experiment with AI tools and providing them with opportunities to contribute to AI projects can cultivate a sense of ownership and enthusiasm for the transformative potential of AI.

The diversity of AI applications and the complexity of integrating them across multiple business functions can create challenges in maintaining cohesion and consistency. As AI initiatives proliferate, ensuring they complement rather than compete with one another is paramount. Establishing a centralized governance framework that oversees AI implementation across the organization can help maintain alignment and coherence. This governance structure should facilitate cross-functional collaboration and communication, ensuring that AI initiatives are integrated seamlessly into your organization's operations and contribute to a unified strategic vision.

To provide a tangible illustration of these challenges, consider the case of a leading financial institution that embarked on an ambitious AI transformation initiative. Despite initial enthusiasm, the project encountered numerous obstacles, including outdated infrastructure, data privacy concerns, and resistance from key stakeholders. By addressing these challenges head-on and implementing a comprehensive AI strategy, the institution eliminated these hurdles and realized AI's full potential. This case study serves as a testament to the resilience and adaptability required to successfully navigate the complexities of AI integration.

Interactive Element: Reflective Journaling Prompt

Take a moment to reflect on the potential integration challenges that may arise in your own organization. Consider the following prompts to guide your reflection:

1.   What challenges might your organization face when integrating AI, and how can you proactively address them?

2.   How can you ensure that your AI initiatives align with your organization's strategic objectives and values?

3.   What steps can you take to foster a culture of openness and collaboration, encouraging employees to embrace AI as a tool for innovation and growth?

By contemplating these questions, you can gain valuable insights into the potential pitfalls of AI integration and develop strategies to navigate them effectively.

2. Lessons Learned

As you stand on the precipice of integrating AI into your business, the allure of cutting-edge technology is undeniable. However, ensuring that the AI tool you choose aligns seamlessly with your organizational goals is paramount.

2.1 Tool Choice

Begin by identifying a clear business problem that the AI needs to solve; that is, which workflows have pain points that need to be improved. This might seem elementary, yet many enterprises fall into the trap of adopting technology for its novelty rather than its necessity. The AI tool should possess the requisite features to address the specified problem effectively, sidestepping the allure of unnecessary bells and whistles that might complicate integration without adding value.

Additionally, the tool should be user-friendly, demanding minimal technical expertise from your team, thus democratizing its use across the organization. Equally crucial is its ability to integrate with existing tools, preventing silos and ensuring a cohesive digital ecosystem. Financial prudence demands that the AI solution is affordable and promises a solid return on investment. In the age of data-driven decisions, the tool must be secure, unbiased, and compliant with legal standards, safeguarding against regulatory repercussions and ethical pitfalls. Before making any substantial commitments, conduct a small-scale pilot to test the solution's viability and efficacy in a controlled environment. This step is indispensable in uncovering potential issues and ensuring that employees and customers find the tool invaluable.

The risk of knowledge loss looms large when engaging external consultants to aid in AI implementation. Often, these experts possess specialized insights that, if not correctly transferred, can leave your organization in a lurch once their contract ends. To mitigate this, it is imperative to contractually require knowledge transfer, encompassing comprehensive handover training and meticulous documentation. This documentation should extend beyond technical specifics to include best practices, troubleshooting steps, and strategic insights that underpin the AI's role within your organization. Appoint an internal AI champion who oversees the implementation process and spearheads future adjustments and optimizations. This individual acts as the custodian of AI knowledge, ensuring continuity and fostering a culture of innovation. In parallel, consider the adoption of no-code or low-code AI tools. These platforms democratize AI capabilities, enabling team members with varying technical skills to contribute without heavy reliance on external developers.

Furthermore, select vendors known for their robust customer support and training programs, ensuring long-term assistance and fostering a partnership rather than a mere transactional relationship. In tandem with these measures, invest in the upskilling of your employees. Equipping your workforce with the skills needed to maintain and optimize AI tools independently instills a sense of ownership and resilience. Complement these efforts with a long-term AI maintenance plan, preemptively addressing potential system breakdowns and ensuring sustained functionality and innovation.

2.2   Short Courses

The common pitfall of quick training programs, hastily organized and without context, often results in ill-prepared employees to wield AI tools effectively. To circumvent this, shift the focus of training from theoretical musings to problem-solving paradigms. Training sessions should be immersive, allowing employees to engage directly with AI tools rather than passively observing tutorials. Include real-world business scenarios, enabling employees to navigate challenges they will likely encounter daily.

Progressively teach AI skills, starting with basic automation before advancing to more complex applications, ensuring a solid foundation before tackling intricate tasks. Training should be continuous, with regular refreshers and hands-on workshops that keep skills sharp and knowledge up-to-date. Utilize your organization's actual data and tasks during training to impart relevance and practicality, transforming abstract lessons into tangible skills.

2.3 AI Champion

Appointing an AI champion within the company provides employees with a go-to resource who can guide, mentor, and troubleshoot as they adapt to the new tools. Encourage employees to pursue business-specific AI certifications, steering clear of generic courses that may not address your organization's unique challenges and opportunities.

2.4 Business-Specific Training

When self-study programs yield lackluster results, seeking industry-specific training avenues is the solution. Professional associations within your industry can be gold mines of targeted training, offering webinars, certifications, and resources tailored to your sector's nuances. Trade conferences and workshops provide hands-on sessions that facilitate learning by doing and invaluable networking opportunities with industry peers. Local business groups and chambers of commerce can also be valuable resources, often offering access to relevant training or mentorship programs that hone in on specific industry challenges. When searching for AI training, prioritize courses and resources focusing on real-world business use cases.

For example, AI-driven ad strategies and customer segmentation insights can revolutionize marketing campaigns. In sales, AI-based lead scoring and chatbots for sales inquiries streamline customer interactions and optimize conversion rates. Process automation and AI-driven inventory management can streamline operations, while customer service can benefit from AI-powered chatbots and email automation that enhance response times and customer satisfaction.

For business-specific AI training, several platforms offer tailored solutions.

- Google's AI for Business provides marketing and analytics use cases

- Microsoft's AI Business School provides strategies tailored to small businesses

- Udacity's AI for Business Leaders focuses on practical AI adoption

- Coursera's AI for Everyone, led by Andrew Ng, explores AI's impact on business strategy

Finding an AI mentor can also be transformative.

- Platforms like Total AI, Clarity.fm, and Upwork connect you with AI specialists who can provide personalized guidance and insights.

- Small Business Development Centers (SBDCs) often offer AI technology mentors who understand smaller enterprises' unique challenges.

- AI accelerators and incubators, such as Techstars AI and Google Startup Accelerator, provide consulting and support to help businesses navigate their AI journey.

When selecting AI tools, prioritize those that offer training and support as part of their package.

- HubSpot AI, for instance, provides AI-powered automation with step-by-step tutorials

- Zapier AI offers guides on automating tasks

- Google Cloud AI delivers business-focused training courses

- OpenAI API is supported by extensive documentation and an active developer community, ensuring you have the resources needed to maximize your AI tools' potential.

2.5   Outsourcing AI Initiatives

Outsourcing AI initiatives can sometimes lead to solutions that neither align with your business objectives nor address the intended problem. To avoid these problems, it's critical to clarify the specific business challenge that AI intends to solve.

- Define what success looks like, whether it's a 30% reduction in response times, a 20% decrease in costs, or a 50% reduction in manual tasks

- Define the measures for the AI system, setting clear KPIs and performance benchmarks to evaluate progress and impact

- Designate a liaison to act as the bridge between your business and the AI vendor. This individual should have a basic understanding of AI to evaluate solutions, regularly review deliverables to ensure alignment with business needs, and provide ongoing feedback rather than wait for the project to conclude.

2.6   Major Milestones

Setting milestones for obtaining intermediate insights throughout the development process will give the stakeholders confidence that the work is staying on track. Begin with business requirement gathering, identifying problems, goals, and success metrics. Move to AI prototype development, incorporating business reviews and feedback loops to refine the solution. Conduct pilot testing in real business scenarios to assess the AI's effectiveness and gather user feedback. Based on this feedback, refine the solution before fully deploying and monitoring.

2.7   Vendor Protocols

When working with vendors, specific deliverables are required including:

- Technical documentation outlining the architecture

- Model training process

- APIs used in the system design

- Business process documentation detailing how AI integrates into workflows

- User guides and training materials showing how to operate and maintain the system

- AI tuning and maintenance guides offer insights into updating and improving the AI system over time, ensuring long-term sustainability

Maintaining a regular vendor feedback cycle is essential for continuous improvement and alignment. Weekly or bi-weekly progress updates are required to stay informed and proactive. Conduct frequent prototype testing to identify and address issues early, assigning test users from different departments to provide diverse, real-world feedback. Live demo reviews ensure the AI functions as intended, allowing for adjustments and optimizations. In the development cycle, consider the industry's specific processes and business needs that the AI is designed to address. Incorporate real-world data from your business to train AI models, ensuring relevance and accuracy. Engage business stakeholders, not just IT teams, to provide comprehensive insights and perspectives.

Finally, when crafting your AI contract, include a trial period to assess the AI's performance, with clear success criteria for transitioning from prototype to full deployment. Establish an exit strategy that ensures all AI data, models, and documentation are transferred to your organization if the vendor terminates the contract. This strategic foresight safeguards your interests and ensures you retain control over your AI initiatives. By following these lessons learned in the field, you can confidently navigate the complexities of AI integration, aligning technology with your business objectives to drive innovation and success.

Chapter Ten:
Emerging Trends In AI And LLMs

"We're making this analogy that AI is the new electricity.
Electricity transformed industries: agriculture,
transportation, communication, and manufacturing."

– Andrew Ng

Envision the busy streets of a city where autonomous vehicles glide silently, seamlessly integrating into the urban hub, while digital billboards adapt in real time to the whims of passersby. This image is not a scene from a distant future but a burgeoning reality shaped by the transformative forces of Artificial Intelligence (AI).

As a business leader, you are at the helm of an organization faced with the imperative of integrating AI's rapid advancements to maintain a competitive edge. The emergence of multimodal AI models, which integrate text, image, and even auditory data, is a testament to the field's expansive potential. These models, capable of processing and synthesizing diverse data types, offer unparalleled opportunities for enhancing customer experiences and personalizing interactions. Imagine a retail platform that recommends products based on past purchases and tailors suggestions based on visual preferences gleaned from uploaded images. Such innovations reshape consumer engagement, offering businesses a sophisticated toolkit for delivering highly personalized experiences.

1. Predicting Industry Shifts: The Future of LLMs

In tandem with these technological marvels, LLMs are evolving with remarkable agility. The advent of few-shot and zero-shot learning has significantly expanded the scope of tasks these models can undertake, enabling them to execute complex functions without extensive retraining. Few-shot learning, for instance, allows LLMs to perform new tasks by being exposed to only a handful of examples, thereby reducing the need for vast amounts of labeled data. This capability is particularly transformative in domains where data scarcity has historically posed a barrier. The integration of LLMs with real-time data streams further amplifies their utility, enabling businesses to derive insights with immediacy, thus informing agile decision-making processes. This fusion of advanced learning techniques and instantaneous data processing sets the stage for a new era of AI applications across industries.

As AI technologies proliferate, the importance of ethical governance and regulation has become increasingly pronounced. The development of international AI governance standards, as discussed in reports such as Stanford's AI Index, reflects a collective acknowledgment of the need for frameworks that ensure AI's responsible deployment. These standards aim to harmonize efforts across borders, fostering a global ecosystem that maximizes AI's benefits and mitigates its risks. Central to this regulatory evolution is the emphasis on transparency and explainability, ensuring that AI systems are not opaque black boxes but tools whose operations can be understood and scrutinized. This transparency is critical for building trust with stakeholders, from customers to regulators, who are increasingly cognizant of the ethical implications of AI.

In parallel, the movement toward decentralized AI systems is gaining momentum, offering a paradigm shift in how AI is developed and deployed. Decentralized AI, which leverages distributed networks and often incorporates blockchain technology, promises enhanced privacy, security, and scalability. These systems mitigate the vulnerabilities associated with centralized data repositories by decentralizing the data and processing tasks, offering a more resilient architecture. Federated learning, a cornerstone of decentralized AI, enables models to be trained across multiple devices without aggregating data on a central server, thereby enhancing data privacy. This approach is particularly compelling in healthcare industries where data sensitivity is paramount. Edge computing, another facet of this trend, brings AI processing closer to the data source, reducing latency and bandwidth usage and facilitating real-time analytics.

2. Predicting Industry Shifts: Case Studies

Imagine the sterile corridors of a hospital transformed by the quiet hum of LLM-powered diagnostics, where doctors wield AI tools to predict patient outcomes with unprecedented accuracy. In healthcare, LLMs are poised to revolutionize diagnostics and patient care by analyzing vast datasets, such as medical journals and patient histories, to offer granular and comprehensive insights. These models can identify subtle patterns and correlations, enabling early detection of diseases that might otherwise remain elusive. As they become integrated into clinical workflows, LLMs will assist physicians in making more informed decisions and tailoring treatment plans to individual patients. This level of precision medicine promises to enhance patient outcomes and reduce healthcare costs, representing a paradigm shift in how we approach medical care.

Imagine a banking system where AI continuously monitors for suspicious patterns, alerting analysts to potential risks before they escalate into crises. Such capabilities enhance security and streamline compliance efforts, providing institutions with a robust framework to navigate the complexities of modern finance. Furthermore, LLMs can assist in developing personalized financial advice and tailoring recommendations to individual client profiles, thus fostering deeper customer engagement and trust.

Retail, too, stands on the brink of transformation as LLMs drive hyper-personalized shopping experiences. By analyzing consumer behavior and preferences, these models can predict what customers want before they even know it themselves. Picture a virtual assistant that curates a shopping list tailored to your tastes, suggesting products based on past purchases and current trends. This level of personalization not only enhances customer satisfaction but also boosts sales by aligning offerings with consumer desires. Retailers leveraging LLMs can optimize inventory management, ensuring that they stock items most likely to appeal to their clientele, reducing waste and increasing profitability.

Beyond specific industries, LLMs are instrumental in the broader digital transformation sweeping across sectors. They automate complex workflows, freeing human capital to focus on strategic initiatives. Integrating LLMs into enterprise resource planning (ERP) systems streamlines operations, enhances data analysis, and improves decision-making processes. This integration allows organizations to harness the power of AI at scale, turning data into actionable insights and driving efficiency across all facets of business operations.

The potential for LLMs to create entirely new business models is vast. Consider AI-driven subscription services that offer real-time insights into consumer behavior, market trends, or operational efficiencies, providing businesses with a competitive edge. Platforms that leverage LLMs for on-demand content creation can revolutionize industries such as media and publishing, enabling rapid, customized content generation at scale. These innovations open new revenue streams and redefine traditional business paradigms, allowing unprecedented growth and flexibility.

While some industries have already begun to capitalize on these advancements, others remain ripe for exploration. In agriculture, LLMs offer the promise of precision farming, where AI analyzes soil conditions, weather patterns, and crop health to optimize yield and sustainability. Legal professionals can streamline contract analysis and document review, reducing the time and cost of complex legal proceedings. These untapped areas represent fertile ground for LLMs to extend their influence, unlocking efficiencies and capabilities that were once the realm of science fiction.

3. Anticipating Challenges: Preparing for the Unknown

As you begin to integrate Large Language Models (LLMs) into your business, one of the foremost challenges lies in managing data privacy within increasingly complex AI ecosystems. The sophistication of LLMs often necessitates access to extensive datasets, which, while enhancing model efficacy, also raises significant privacy concerns. In an era where data breaches are not merely hypothetical risks but tangible threats with severe repercussions, safeguarding sensitive information becomes paramount. Ensuring robust data protection measures and compliance with regulations such as the General Data Protection Regulation (GDPR) is crucial. However, the complexity of AI systems and the sheer volume of data they handle can complicate these efforts, necessitating a nuanced approach to privacy management that balances the need for data access with stringent security protocols.

Beyond privacy, ensuring the robustness and reliability of LLMs across diverse environments presents another formidable challenge. The efficacy of these models can vary significantly depending on the context in which they are deployed, influenced by factors such as data quality, computational resources, and the specific nuances of the application domain. It's imperative to implement rigorous testing protocols and validation procedures that account for variability in data and operational conditions to mitigate these challenges. This task involves fine-tuning models to enhance performance and establishing contingency plans to address potential failures or inaccuracies. By fostering a culture of resilience, where adaptability and continuous improvement are integral to the organizational ethos, businesses can better navigate the complexities of LLM deployment.

In the face of uncertainty, scenario planning and risk assessment frameworks become invaluable tools for managing the inherent risks associated with emerging LLM technologies. Developing a comprehensive understanding of potential risks and their implications allows businesses to craft adaptive strategies that facilitate agile responses to market changes. This proactive approach enables organizations to pivot swiftly and effectively, ensuring that they remain competitive and resilient in the face of evolving technological innovations. Moreover, by cultivating a culture of flexibility and adaptability, where continuous learning and innovation are not only encouraged but institutionalized, businesses can enhance their capacity to anticipate and respond to unforeseen challenges.

As LLMs become more widely used, the regulatory and ethical standards surrounding their use will likely become increasingly complex. Implementing international AI compliance requirements requires a keen understanding of the legal and ethical considerations that underpin AI deployment. This task includes addressing ethical concerns related to AI decision-making, such as potential biases and fairness issues, which can have significant implications for stakeholder trust and organizational reputation. Ensuring that AI systems are transparent, accountable, and aligned with ethical standards necessitates a concerted effort to engage with regulatory bodies, industry experts, and other stakeholders. By fostering a culture of ethical governance and accountability, businesses can mitigate potential regulatory hurdles and build trust with customers, partners, and the broader community.

4. Exploring Disruptive Innovations: LLMs as Game Changers

In customer engagement, LLMs have fundamentally transformed how businesses interact with their clientele by ushering in a new era of AI-powered chatbots that redefine customer service. These sophisticated virtual assistants, capable of understanding and responding to inquiries with human-like nuance, have increased efficiency by handling routine queries, and have elevated the customer experience by providing instant, accurate, and personalized responses. This transformation allows businesses to maintain a 24/7 presence, ensuring that customers receive the support they need anytime, thus fostering loyalty and enhancing brand reputation. In industries where customer interaction is paramount, such as retail or hospitality, the shift from human-operated to AI-driven communication channels is not just an operational change but a strategic reorientation that can significantly impact the bottom line.

The disruption extends beyond customer service into the creative areas of content generation, where LLMs are revolutionizing how content is conceptualized and produced. The emergence of automated content creation tools powered by LLMs offers businesses an unprecedented capacity to generate marketing copy, articles, and even creative narratives at scale. These tools can analyze vast amounts of existing content, identify trends, and produce text that mirrors the tone and style desired, thereby reducing the time and cost traditionally associated with content development. This technology streamlines operations for media companies and marketing agencies and enables rapid adaptation to consumer preferences, ensuring timely and relevant content delivery.

LLMs offer a competitive edge in product development by providing AI-driven insights that can significantly enhance innovation cycles. By analyzing consumer feedback, market trends, and historical data, LLMs can identify gaps and opportunities, guiding the development of products that align with customer needs and expectations. This data-driven approach reduces the uncertainty inherent in product innovation, allowing companies to make informed decisions with a higher likelihood of success. Similarly, in supply chain operations, predictive analytics enabled by LLMs play a pivotal role in optimizing logistics and inventory management. By forecasting demand and identifying potential disruptions, businesses can streamline operations, minimize costs, and improve service delivery, thereby gaining a distinct advantage in a competitive market.

As LLMs reshape industries, they also redefine workforce dynamics, necessitating a shift in job roles and skill sets. AI specialists possess the expertise to develop, implement, and manage these sophisticated systems. The role is creating new career opportunities while transforming existing roles. Employees are increasingly required to engage in continuous learning and reskilling initiatives to remain relevant in an AI-driven workplace. This evolution encourages a culture of lifelong learning, where adaptability and technical proficiency become invaluable assets.

Investing in workforce development ensures organizations can fully leverage AI technologies, maximizing their potential to drive innovation and growth.

The transformative power of LLMs also unlocks opportunities for innovation and collaboration, paving the way for co-creation partnerships between companies and AI providers. These alliances enable businesses to access cutting-edge technologies and expertise, facilitating the development of bespoke AI solutions tailored to specific needs. Joint ventures exploring new AI-driven markets are becoming more prevalent as companies recognize the potential for LLMs to open new revenue streams and business models. By fostering an ecosystem of collaboration, businesses can accelerate innovation, share risks, and capitalize on emerging opportunities, positioning themselves at the forefront of the AI revolution.

5. Continuous Learning: Keeping Pace with AI Advancements

In the relentless march of technological progress, the imperative for continuous learning has never been more critical. As AI evolves at an unprecedented pace, the need for businesses and individuals to adapt through lifelong learning initiatives is paramount. AI literacy is no longer a luxury but a necessity for staying relevant in today's competitive marketplace. By incorporating AI topics into corporate training programs, organizations can equip their workforce with the knowledge and skills required to navigate this complex terrain. This approach fosters a learning culture and empowers employees to engage with AI technologies confidently, ensuring they remain valuable contributors to the organization's success.

Creating a continuous learning and innovation culture requires more than just formal training programs. It involves encouraging experimentation and exploration of new AI technologies, allowing individuals to push boundaries and discover novel solutions. By fostering an environment where curiosity is rewarded, businesses can unleash the creative potential of their teams and drive transformative change. A knowledge-sharing ecosystem within the organization further amplifies this effect, enabling employees to exchange ideas, insights, and experiences. This collaborative approach accelerates learning and cultivates a sense of community and shared purpose, driving collective growth and progress.

The role of technology in delivering scalable and accessible learning experiences cannot be overstated. Online platforms offering AI courses and certifications provide flexible learning opportunities for learners to tailor to their needs and schedules. These platforms democratize access to education, enabling employees at all levels to upgrade their skills and knowledge. AI-driven learning personalization tools enhance this process by adapting content to the individual's learning style and pace. This personalized approach improves engagement and retention and ensures each learner receives the most relevant and impactful education, maximizing the return on investment in training initiatives.

Collaboration and knowledge sharing are essential for driving collective learning and growth. Interdepartmental AI learning initiatives break down silos and foster cross-functional collaboration, enabling teams to leverage diverse perspectives and expertise. This multidisciplinary approach enhances problem-solving and drives innovation by combining insights from various domains. External collaborations with academia and industry experts further enrich this process, providing access to cutting-edge research, emerging trends, and best practices. By engaging with a broader ecosystem of knowledge, organizations can stay ahead of the curve and remain competitive in an increasingly complex and dynamic area.

Interactive Element: Reflection Section

Consider how your organization currently approaches learning and development. Reflect on the following questions to identify opportunities for fostering a culture of continuous learning:

1. How does your organization incorporate AI topics into its training programs?

2. What strategies are in place to encourage experimentation and knowledge sharing?

3. How can technology be leveraged to enhance learning experiences and engagement?

4. What opportunities exist for collaboration with external partners in academia and industry?

By addressing these questions, you can gain valuable insights into your organization's learning culture and identify actionable steps to enhance it. As AI continues to evolve, fostering a culture of continuous learning and innovation will be crucial for staying competitive and unlocking the full potential of AI technologies.

6. The Road Ahead: A Vision for AI-Driven Business Transformation

Imagine a future where AI is not merely a tool but a fundamental component of your business's strategic core, seamlessly integrated into every facet of operations. This vision of AI-driven transformation heralds a new era in which AI, particularly Large Language Models (LLMs), drives sustainable innovation and growth across industries. AI will become an indispensable ally as businesses evolve, offering insights that propel decision-making and foster a culture of agility and innovation. AI will anchor strategic initiatives no longer confined to the peripheries, enabling organizations to anticipate and adapt to market changes with unprecedented speed and precision.

As we gaze toward this horizon, the potential of AI to address global challenges becomes increasingly apparent. In environmental sustainability, AI-driven solutions are ready to play a pivotal role in mitigating climate change, optimizing energy consumption, and enhancing resource management. Imagine AI algorithms that predict environmental trends, enabling proactive measures to reduce carbon footprints and bolster sustainability efforts. Beyond environmental applications, AI's impact on global healthcare access is equally transformative. By leveraging AI technologies, healthcare providers can extend their reach to underserved populations, offering diagnostic tools and treatment plans tailored to individual needs. This democratization of healthcare promises to bridge disparities, ensuring that quality care is accessible to all, regardless of geographic or economic barriers.

When working through these promising innovations, the importance of ethical and responsible AI development cannot be overstated. A commitment to transparency and accountability in AI practices is crucial for building stakeholder trust and ensuring positive outcomes. Businesses must prioritize ethical governance, embedding principles of fairness and inclusivity into their AI strategies. By fostering an environment of openness and integrity, organizations comply with emerging regulatory standards and align with the values of their customers, partners, and communities. This alignment fosters a culture of trust where stakeholders are confident in the ethical deployment of AI technologies.

Businesses must identify and pursue key milestones that guide their progress in achieving a successful AI-driven transformation. Achieving AI maturity within the organization is crucial, signifying the seamless integration of AI technologies across functions and processes. This maturity level positions AI as a core competency, offering a competitive advantage that enables businesses to innovate and thrive in a rapidly changing market. Establishing AI as a foundational element of business operations requires a strategic commitment to continuous improvement, ensuring that AI systems evolve in tandem with organizational goals and market dynamics.

As AI continues to permeate industries, it will redefine the contours of business, offering new pathways for growth and innovation. The journey toward AI-driven transformation is rich with potential, inviting firms to explore the limitless possibilities of AI while remaining grounded in principles of ethical governance and responsible development. This transformative era calls for bold vision and strategic foresight, empowering organizations to harness the power of AI for the greater good. As we stand at the precipice of this exciting future, the road ahead is illuminated by the potential of AI to shape a more interconnected, efficient, and sustainable world.

7. AI Capabilities in 2035: A Ten-Year Projection

There will be AI models to process information orders of magnitude faster than today while consuming substantially less energy. Quantum-classical hybrid computing may become mainstream for specific AI applications, particularly in optimization problems and complex simulations.

Multimodal AI will reach maturity, seamlessly integrating text, voice, image, video, and sensor data to create more comprehensive understanding and responses. These systems will process real-world complexity more naturally, understanding context across multiple dimensions simultaneously rather than requiring separate processing pathways for different input types.

AI reasoning capabilities will likely evolve beyond current pattern matching to demonstrate more sophisticated logical reasoning, causal understanding, and creative problem-solving. Systems may exhibit improved ability to work with incomplete information, make reasonable assumptions, and explain their reasoning processes in ways humans can understand and verify.

Long-term planning and goal-oriented behavior will become more robust, with AI systems capable of pursuing complex objectives over extended time frames while adapting to changing circumstances. This includes better understanding of consequences, trade-offs, and ethical implications of decisions.

Physical AI embodiment will advance significantly, with robots and autonomous systems demonstrating much greater dexterity, mobility, and environmental awareness. Manufacturing, logistics, and service robotics will become more prevalent as costs decrease and capabilities increase. Autonomous vehicles will likely achieve broader deployment, though the timeline may vary significantly by region and use case.

By 2035, AI systems will likely demonstrate significantly enhanced computational efficiency through advances in specialized hardware and algorithms. We can expect

AI systems will demonstrate improved ability to operate in unstructured, unpredictable environments rather than requiring highly controlled conditions. This includes better handling of edge cases and novel situations that had not been explicitly programmed or included in the latest training .

AI creative capabilities will likely expand beyond current text and image generation to include sophisticated video production, interactive media creation, and potentially novel forms of digital art and entertainment. These systems may collaborate with humans on complex creative projects, contributing original ideas while respecting human creative vision and intent.

Scientific research acceleration through AI will become more pronounced, with systems capable of generating and testing hypotheses, analyzing complex datasets, and even conducting certain types of experiments autonomously. This may significantly speed up discovery processes in fields like drug development, materials science, and climate research.

AI systems will demonstrate much more sophisticated personalization, adapting not just to stated preferences but to observed behavior patterns, changing needs, and life circumstances. Personal AI assistants may maintain long-term relationships with users, understanding their goals, values, and decision-making patterns.

Contextual awareness will reach new levels, with AI systems understanding not just immediate context but broader situational, cultural, and temporal factors that influence appropriate responses and recommendations.

Despite these advances, significant challenges will likely persist. AI systems may still struggle with true understanding versus sophisticated pattern matching, though the distinction may become less practically relevant. Issues of bias, fairness, and transparency will require ongoing attention and may become more complex as systems become more sophisticated.

Energy consumption and computational requirements may continue to be constraining factors, even with efficiency improvements. The need for human oversight, especially in high-stakes decisions, will likely remain important regardless of AI capabilities.

By 2035, AI will likely be deeply integrated into daily life and business operations, though this integration may be less visible as systems become more seamless and natural to interact with. The economic impact will be substantial, with new job categories emerging while others evolve or become automated.

Regulatory frameworks and governance structures for AI will mature significantly, providing clearer guidelines for development and deployment while balancing innovation with safety and ethical considerations.

The next decade will likely see AI transition from today's impressive but narrow applications to more general-purpose, collaborative systems that augment human capabilities across a wide range of domains. For someone starting their career today, this represents an enormous opportunity to build expertise in emerging technologies while developing the uniquely human skills that will remain valuable in an AI-enhanced world.

The key insight here is that the most successful approaches to monetizing AI will likely involve human-AI collaboration rather than replacement, focusing on how AI can amplify human creativity, judgment, and relationship-building capabilities rather than simply automating tasks.

Conclusion

As we conclude our exploration of Large Language Models (LLMs) and their transformative potential for business, let us take a moment to reflect on the remarkable journey we have embarked upon together. From the theoretical foundations that sparked the AI revolution to the practical applications that are reshaping industries, LLMs have emerged as a catalyst for change, challenging traditional paradigms and opening up new frontiers of innovation.

Throughout this book, we have sought to demystify the complex world of AI. By breaking down technical jargon into accessible concepts, we hope that you, the business leader, can confidently engage in AI discussions and make informed decisions. By simplifying the language of AI, we have aimed to bridge the gap between technology and strategy, enabling you to harness the full potential of LLMs in your organization.

Central to this endeavor has been the imperative of calculating and communicating the return on investment (ROI) of LLM initiatives. We have explored various methods for assessing AI's tangible and intangible benefits, equipping you with the tools to craft compelling business cases that resonate with stakeholders across the board. By effectively articulating the value proposition of LLMs, you can rally support for your AI initiatives and secure the resources necessary to drive meaningful change.

As we have seen, the path to AI adoption has its challenges. Balancing innovation with risk management is a delicate dance that requires finesse and foresight. By proactively addressing concerns surrounding data privacy, security, and ethical considerations, you can build trust with your stakeholders and ensure that your AI initiatives are grounded in principles of responsibility and transparency. Developing

129

robust governance frameworks and a commitment to continuous monitoring and improvement are essential for understanding the complexities of AI integration.

Equally crucial is the cultivation of a future-ready workforce, one that is ready to thrive in an AI-driven world. We have emphasized the importance of identifying new roles, upskilling existing talent, and fostering a culture of continuous learning and innovation. By investing in your human capital and nurturing an environment where workers can satisfy their curiosity through experimentation, you can build the resilience necessary to adapt to the ever-evolving demands of the AI era.

The strategic implementation of LLMs is a multifaceted endeavor that requires careful planning and execution. We have walked through the steps for integrating LLMs into your business processes, from identifying key use cases to piloting initiatives and scaling successes. The real-world case studies peppered throughout this book testify to the transformative power of LLMs, offering valuable lessons and best practices for overcoming common adoption challenges.

As we gaze into the future, the potential of LLMs to reshape industries is both exhilarating and humbling. From healthcare and finance to retail and beyond, the applications of LLMs are limited only by our imagination. By staying attuned to emerging trends and anticipating industry shifts, you can position your organization at the forefront of the AI revolution, ready to seize the opportunities that lie ahead.

The journey toward AI-driven business transformation is not a solitary one. It requires the collective effort of visionary leaders, skilled technologists, and engaged stakeholders, all working together toward a shared vision. By aligning your AI initiatives with your core business objectives, you can ensure that your investments yield tangible results and drive sustainable growth.

And so, I invite you to embrace the insights and strategies we have explored together, to apply them to your unique business context, and to lead your organization into a future shaped by the boundless potential of AI. The road ahead may be uncharted, but with the right mindset, the right tools, and the right partners, you have the power to navigate it with confidence and grace.

As you embark on this exhilarating journey, remember that the accurate measure of success lies not in the algorithms or the models but in the people, you touch and the value you create. May you lead with courage, innovate with purpose, and transform your business for the better.

The future is yours to shape. Let us forge ahead, together, into an AI-driven world of limitless possibilities.

Bibliography

AIM Research. (2024, May 22) *10 generative AI success stories: How businesses transformed their operations.* https://aimresearch.co/market-industry/10-generative-ai-success-stories-how-businesses-transformed-their-operations

AIHA. (2022, February 15). *Effective communications for artificial intelligence projects.* https://www.aiha.org/blog/effective-communications-for-artificial-intelligence-projects

ARM. (n.d.). *Customer Success Stories.* https://www.arm.com/company/success-library

AWS Training and Certification. (2024, November 8). *Upskill your team for generative AI projects with AWS training.* https://aws.amazon.com/blogs/training-and-certification/upskill-your-team-for-generative-ai-projects-with-aws-training/

Chen, R., et al. (2024, October 11). Google's cloud well-architected framework, AI and ML perspective: Reliability. *Google Cloud.* https://cloud.google.com/architecture/framework/perspectives/ai-ml/reliability

CIO. (2024, July 15). *Best practices for integrating AI in business: A governance approach.* https://www.cio.com/article/2517557/best-practices-for-integrating-ai-in-business-a-governance-approach.html

Cognizant. (2023, November 28). *Strategies for overcoming organizational resistance to AI.* https://www.cognizant.com/nl/en/insights/blog/articles/from-resistance-to-advocacy

Corporate Finance Institute. (n.d.). *Data anonymization: Overview, techniques, advantages.* https://corporatefinanceinstitute.com/resources/business-intelligence/data-anonymization/

Dial Zara. (2024, May 31). *Cross-functional teams for AI success: Guide.* https://dialzara.com/blog/cross-functional-teams-for-ai-success-guide/

English, L. (2023, May 25). The impact of AI on company culture and how to prepare now. *Forbes.* https://www.forbes.com/sites/larryenglish/2023/05/25/the-impact-of-ai-on-company-culture-and-how-to-prepare-now/

Forbes Business Development Council. (2024, September 6). 20 leadership skills that are still relevant in the AI age. *Forbes.* https://www.forbes.com/councils/forbesbusinessdevelopmentcouncil/2024/09/06/20-leadership-skills-that-are-still-relevant-in-the-ai-age/

Forbes. (2023, October 25). 10 hurdles companies are facing when implementing AI and how to overcome them. https://www.forbes.com/councils/theyec/2023/10/25/10-hurdles-companies-are-facing-when-implementing-ai-and-how-to-overcome-them/

Google for Developers. (2005, June 2). Machine learning crash course: Classification. https://developers.google.com/machine-learning/crash-course/classification/roc-and-auc

Google for Developers. (2025, May 22). Machine learning crash course: Linear regression. https://developers.google.com/machine-learning/crash-course/linear-regression/loss?hl=en

GSMA AI for Impact Toolkit. (n.d.). *Governance model for ethical AI.* https://aiforimpacttoolkit.gsma.com/responsible-ai/governance-model-for-ethical-ai/

IEEE Xplore. (2024, October 16). *Impact of transformer-based models in NLP: An in-depth analysis.* https://ieeexplore.ieee.org/document/10710796/

Innosight. (2023, November 30). *A glossary of common AI terms for business leaders.* https://www.innosight.com/insight/a-glossary-of-common-ai-terms-for-business-leaders/

McKinsey & Company. (2021, April 16). *Building a learning culture that drives business forward.* https://www.mckinsey.com/capabilities/people-and-organizational-performance/our-insights/building-a-learning-culture-that-drives-business-forward

Mitchell, R. K., Agle, B. R., & Wood, D. J. (1997). Toward a theory of stakeholder identification and salience: Defining the principle of who and what really counts. *Academy of Management Review*, 22(4), 853-886.

Moveworks. (n.d.). *AI terms glossary: AI terms to know in 2024.* https://www.moveworks.com/us/en/resources/ai-terms-glossary

NIST AIRC. (n.d.). *AI risks and trustworthiness.* https://airc.nist.gov/AI_RMF_Knowledge_Base/AI_RMF/Foundational_Information/3-sec-characteristics

Orulluoğlu, O. (2023, July 16). R-square in machine learning: A powerful tool for evaluating model performance. *Medium.* https://medium.com/@bayramorkunor/r-square-in-machine-learning-a-powerful-tool-for-evaluating-model-performance-f90b43e23d9b

Partnership on AI. (2024, July 25). *AI needs inclusive stakeholder engagement now more than ever.* https://partnershiponai.org/ai-needs-inclusive-stakeholder-engagement-now-more-than-ever/

Perception Point. (n.d.). *AI security: Risks, frameworks, and best practices.* https://perception-point.io/guides/ai-security/ai-security-risks-frameworks-and-best-practices/

Princeton Dialogues on AI and Ethics. (2025). *Case studies.* https://aiethics.princeton.edu/case-studies/

Sand Technologies. (2025, January 10). *A practical guide to measuring AI ROI.* https://www.sandtech.com/insight/a-practical-guide-to-measuring-ai-roi/

SAS Blogs. (2024, March 14). *The importance of data quality for large language models.* https://blogs.sas.com/content/sascom/2024/03/14/the-importance-of-data-quality-in-using-large-language-models/

Securiti.ai. (2023, September 29). *The impact of the GDPR on artificial intelligence.* https://securiti.ai/impact-of-the-gdpr-on-artificial-intelligence/

Snorkel AI. (2023, May 25). *Large language models: History, pros, cons.* https://snorkel.ai/large-language-models/

Stanford Online. (n.d.). *The 4 steps to building an effective AI strategy.* https://online.stanford.edu/4-steps-building-effective-ai-strategy

Stanford Institute for Human-Centered AI. (2025). *Stanford's AI index report 2025.* https://hai.stanford.edu/ai-index

Structural Learning. (2023, July 20). *Chomsky's theory - Universal grammar.* https://www.structural-learning.com/post/chomskys-theory

Teamed. (2023, December 18). *The AI jobs boom: How AI shaped the workforce in 2023.* https://www.teamed.global/blog/the-ai-jobs-boom-how-ai-shaped-the-workforce-in-2023

Tech-Stack. (2024, August 21). *Measuring the ROI of AI: Key metrics and strategies.* https://tech-stack.com/blog/roi-of-ai/

UNESCO. (n.d.). *Ethics in artificial intelligence.* https://www.unesco.org/en/artificial-intelligence/recommendation-ethics

V2A Consulting. (2023, August 15). *Understanding large language models (LLMs) and their business applications.* https://v2aconsulting.com/insights/understanding-large-language-models-and-their-business-applications/

Version 1. (n.d.). *AI performance metrics: The science & art of measuring AI.* https://www.version1.com/en-us/blog/ai-performance-metrics-the-science-and-art-of-measuring-ai/

Wikipedia. (n.d.). *Transformer (deep learning architecture).* https://en.wikipedia.org/wiki/Transformer_(deep_learning_architecture)

World Economic Forum. (2019, December 2). *Davos manifesto 2020: The universal purpose of a company in the fourth industrial revolution.* https://www.weforum.org/stories/2019/12/davos-manifesto-2020-the-universal-purpose-of-a-company-in-the-fourth-industrial-revolution/

Zahid, Idrees A. et al. (2024). *Unmasking large language models by means of OpenAI. ScienceDirect. Volume 23,* (September). https://www.sciencedirect.com/science/article/pii/S2667305324001054

Zoho Analytics. (2025, April 9). *The 5 best AI data visualization tools for 2025.* https://www.zoho.com/analytics/insightshq/5-best-ai-data-visualization-tools.html